An Introduction to Robophilosophy
Cognition, Intelligence, Autonomy, Consciousness, Conscience, and Ethics

RIVER PUBLISHERS SERIES IN AUTOMATION, CONTROL AND ROBOTICS

Volume 2

Series Editor

TAREK SOBH
University of Bridgeport
USA

ANDRÉ VELTMAN
PIAK and TU Eindhoven
The Netherlands

The "River Publishers Series in Automation, Control and Robotics" is a series of comprehensive academic and professional books which focus on the theory and applications of automation, control and robotics. The series focuses on topics ranging from the theory and use of control systems, automation engineering, robotics and intelligent machines.

Books published in the series include research monographs, edited volumes, handbooks and textbooks. The books provide professionals, researchers, educators, and advanced students in the field with an invaluable insight into the latest research and developments.

Topics covered in the series include, but are by no means restricted to the following:

- Robots and Intelligent Machines
- Robotics
- Control Systems
- Control Theory
- Automation Engineering

For a list of other books in this series, visit www.riverpublishers.com

An Introduction to Robophilosophy
Cognition, Intelligence, Autonomy, Consciousness, Conscience, and Ethics

Spyros G. Tzafestas

National Technical University of Athens
Athens, Greece

River Publishers

Published, sold and distributed by:
River Publishers
Alsbjergvej 10
9260 Gistrup
Denmark

River Publishers
Lange Geer 44
2611 PW Delft
The Netherlands

Tel.: +45369953197
www.riverpublishers.com

ISBN: 978-87-93379-57-2 (Hardback)
       978-87-93379-56-5 (Ebook)

Man, a being in search of meaning.
Science is the creator of prosperity.

Plato
Athens, 428–348 B.C.

Contents

*Philosophy is the peculiarly stubborn
attempt to think clearly.*
William James

*In law a man is guilty when he violates the rights
of others, in ethics he is guilty if he only thinks of doing so.*
Immanuel Kant

*To be useful, a machine has to do more
than just correctly perform some task.*
John McDermott

Preface

This book is concerned with *robophilosophy* (robot philosophy), a term introduced in the 2014 Robophilosophy Conference (Aarchus University, Joanna Seibt; www.conferences.au.dk/robo-philosophy). The book presents the fundamental concepts and philosophical aspects of the six principal mental capabilities of modern robots, namely: cognition, intelligence, autonomy, consciousness, conscience, and ethics. For the reader's convenience, the book begins with a short introduction to the basic issues of general philosophy, philosophy of science, and philosophy of technology at an elementary tutorial level. Modern robots go beyond the factory and industrial shop floor, embracing societal and everyday tasks such as home services, mobility assistive tasks, medical functions, entertainment, emotional attachment and elderly companionship, also including war processes. The two dominant categories of modern mental robots are the anthropomorphic (humanoid) robots and the zoomorphic (animaloid) robots. These robots are able to socially interact with humans, exhibiting behaviors similar to human beings or animals.

Cognitive robots cannot be defined in a unique formal way, but they are defined through their capabilities. They actually represent the product of technological embodiment of cognitive features to robots. Robot intelligence enables robots to solve problems under uncertain or varying conditions of their environment. It can be of several levels, and can be measured by various machine intelligence tests. Autonomy has several meanings, and in robotics it is interpreted as independence of control. It can be implemented in several layers of partial (weak) autonomy between full (strong) autonomy and non-autonomy. Consciousness can be interpreted as the mind or the mental capacities exhibited via thoughts, feelings, and volition. Four meanings of consciousness are: consciousness as perceiving the world, consciousness as being awake, consciousness as being unique, and consciousness as observing its reasoning. Consciousness in robots can be achieved by embodying to them a number of specific cognitive features. Conscience is the inner mechanism that warns humans when their actions are in disagreement with their own values, norms, and beliefs. Ethics deals with the moral beliefs and rules, and, actually, it is the implementation of conscience in an organized way. Robot ethics involves the ethics of people who construct and employ robots, and the ethics of how humans treat robots.

This book provides a consolidated introductory overview of the philosophical aspects of these mental abilities as applied to robots, also including some technical aspects of their implementation architectures suitable for the student. It is primarily intended for use as a general information source, but it can be profitably used as a complement in robotics courses on the philosophical and humanistic side. Teachers and researchers of robotics, especially those interested in the philosophy of robotics, may find this book a convenient starting point.

Spyros G. Tzafestas
Athens

List of Figures

List of Tables

List of Abbreviations

AA	Alphabetic Arithmetic
AC	Association Cortex
ACS	Action-Centered Subsystem
ACT-R	Adaptive Control of Thought-Rational
ALFUS	Autonomy Levels for Unmanned Systems
Am	Amygdala
BG	Basal Ganglia
BT	Biological Theory
CERA	Conscious and Emotional Reasoning Architecture
CLARION	Connectionist Learning with Adaptive Rule Induction ON-line
CogAff architecture	Cognitive and Affective Architecture
Con Scale	Consciousness Scale
CRCC	Canadian Commission on Rehabilitation Councellor Certification
Cyborg	Cybernetic Organism
EPIC	Executive Process Interactive Control
FOA	Focus of Attention
GWA	Global Workspace Architecture
GWT	Global Workspace Theory
HCMP	Hippocampus
H-CogAff	Human-like CogAff
HOT	Higher-Order Thought Theory
HRI	Human-Robot Interface
IBCA	Integrated Biologically-based Cognitive Architecture
IITC	Information Integrated Theory of Consciousness
IPDI	Principle of Increasing Precision with Decreasing Intelligence
K-CERA	Khepera CERA
LIDA	Learning Intelligent Distribution Agent

MANN	Modular Artificial Neural Network architecture
MC	Motor Cortex
MCS	Meta-Cognitive Subsystem
MS	Motivation Subsystem
NACS	Non-Action-Centered Subsystem
NARS	Non-Axiomatic Reasoning System
NCC	Neural Correlates for Consciousness
PC	Process Control
PFC	Prefrontal Cortex
PMC	Perceptual and Motor Cortex
RESNA	Rehabilitation Engineering Technology Society
ROC	Regions of Certainty
ROU	Regions of Uncertainty or fuzziness
SC	Sensory Cortex
SOAR	State, Operator And Result
SRT	Serial Reaction Time
SS-RICS	Symbolic and Sub-symbolic Robotics Intelligent Control System
SSS	Servo-layer/Subsumption layer/Symbolic layer architecture
TCA	Task Control Architecture
TOH	Tower of Hanoi
ToM	Theory of Mind
WHO	Wrist Hand Orthotic
WM	Working Memory

Gallery of Some Famous Philosophers

Thales   Socrates   Plato   Aristotle Anaximander
624-547 BC 469-399 BC 341-271BC 384-322 BC 610-546 BC
Unit of   Metaphysics Truth &   Metaphysics All is One
Existence Absolute Truth Reality Space & Motion but Infinite

Heraclitus   Parmenides Democritus   Epicurus   Anaximenes
535-475 BC 515-460 BC 460-370 BC 341-271 BC 600-540 BC
All is   Change   Atom Theory Happiness   All is One
Becoming   &Illusion   of Matter   & Desire   & Aether

Marcus   Rene   Leibnitz   I. Kant George Berkley
Aurelius   Descartes 1646-1716 1724-1804 1685-1753
121-180   1596-1650 Metaphysics Pure Reason Idealism
All is One   Rationalism Monadology Deontology Mind & God

David Hume Spinoza John Stuart Mill Bertrand Russel Rousseau
1711-1776   1632-1677 1806-1873   1872-1970   1712-1788
Matter's Wave Purposeful Utilitarianism Analysis of Education
Structure   Action   & Justice   Mind   Philosophy

A Sample of Famous Philosophy Quotes

Plato

Those whose hearts are fixed on Reality itself deserve the title of Philosopher.

Socrates

Only Absolute Truth is that there are No Absolute Truths.

Aristotle

Knowing yourself is the beginning of all wisdom. Nature does nothing uselessly.

Heraclitus

All things come out of the One and the One out of all things.

Albert Einstein

If you can't explain it simply, you don't understand it well enough.

No problem can be solved from the same level of consciousness that created it.

Isaac Newton

Truth is ever to be found in simplicity and not in the multiplicity and confusion of things.

Bertrand Russell

Mathematics is I believe the chief source of the belief in eternal and exact truth as well as a sensible intelligent world.

Ray LeBlond

You learn something every day if you pay attention.

Confucius

I hear and I forget. I see and I remember. I do and I understand. It does not matter how slowly you go so long as you do not stop.

Marcus Aurelius

We should not say 'I am an Athenian' or 'I am a Roman' but 'I am a citizen of the Universe'.

Seneca

Life itself is not a good nor an evil; life is where good or evil find a place depending on how you make it for them.

Leibniz

Indeed in general I hold that there is nothing truer than happiness and nothing happier and sweeter than truth.

Ralph Waldo Emerson

Character is higher than intellect.

St Augustin

We are too weak to discover the truth by reason alone.

Mohandas Gandhi

There is a higher court than courts of justice and that is the court of conscience. It exceeds all other courts.

Abstracts

Chapter 1: Background Concepts and Outline of the Book

Modern robots possess, in addition to mechanical dexterities, a variety of mental capabilities where perception, processing, autonomy, and action are embodied in a human-like (anthropomorphic) or animal-like (zoomorphic) form. Robotics with this kind of features aims to create robot beings to interact with human beings in order to perform sociable functions. Anthropomorphic and zoomorphic robots are expected to change the way humans interact with machines. This chapter provides a brief preparatory account of the primary elements of mental robots (which are the subject of the book), namely: cognition, intelligence, autonomy, consciousness, conscience, and ethics, including the processes of learning and attention. An outline of the book is also provided.

Chapter 2: Philosophy: Fundamental Issues

Philosophy (love of wisdom) is about our thoughts, beliefs and attitudes, about ourselves and the world (Cosmos). Philosophy can be interpreted as the "process of philosophizing" (i.e., doing philosophy) in which linguistic symbol tokens of speaking, hearing, reading, and writing or drawing, are included. Western philosophy has its origin in ancient Greek Philosophy which was concerned with two principal questions: "what is" and "what is being". In this chapter we give an elementary introduction to the fundamental general issues of philosophy, namely: basic features of philosophy, and fields of philosophy (ontology, epistemology, teleology, ethics, and aesthetics). The material of the chapter is focused on the definition, the etymology, and some of the fundamental questions addressed in each field.

Chapter 3: Philosophy of Science and Technology

Philosophy of science is concerned with the understanding of what are experimental methods, observations and scientific theories or models, and how they have enabled scientists to reveal so many secrets of the natural world. Philosophy of technology is actually a broad field crossing many other fields of human activity. Many philosophers argued (not convincingly) that technology is "applied science". Their argument is that technology involves two parts, i.e., "substantial (scientific) theories", and "operative theories". In this chapter we start by discussing 'what is science' and "what is the distinction of science from non-science". Then, we discuss the philosophical principles of rationalism, logic reasoning, empiricism, idealism, positivism, and the general aspects and questions of the philosophy of technology.

Chapter 4: Robot Cognition

In general, cognition is defined as the process of knowing, and includes the aspects of perception, awareness, reasoning, judgment, and intuition. Cognitive science is the science of mind. Historically, cognitive science has emerged with the behaviorism (stimulus-response) approach, and then followed by the so-called cognitive revolution that studies the structure of the internal information processing. In this chapter we discuss several fundamental philosophical aspects and questions of both human cognition and robot cognition. Specifically, the chapter starts with the ontological questions (what is cognition, what is cognitive robotics), followed by basic historical landmarks and a discussion of the philosophy of cognitive robotics. Then, some technological elements of cognitive robots based on the cognivist, the emergent, and the hybrid cognivist/emergent systems approaches are given. Finally, the chapter outlines a set of requirements for the design of cognitive robots.

Chapter 5: Robot Intelligence

Intelligence is the mental capability of humans which involves many complex psychological phenomena. Machine intelligence differs from human intelligence. The most fundamental issue in the area of machine intelligence is whether the machines will ever be able to think like humans. Turing has developed a test for judging whether machines are genuinely thinking. This is known as "Turing test" and has attracted extensive criticism based on the view that computers can only do what we program them to do. This chapter investigates some philosophical aspects of human and machine intelligence focusing on the ontological issues. Then, it summarizes the philosophical views of eminent scientists about machine/robot intelligence, and reviews the Turing machine intelligence test along with several alternative tests formulated over the years.

Chapter 6: Robot Autonomy

Autonomy is a major philosophical and practical issue for human personal and societal life and prosperity. It has several senses and interpretations from the social, moral, and political viewpoints. In public health, autonomy refers to the autonomous decision making of a patient, and in government, autonomy refers to one's own self-governance. In general, rational autonomy refers to an agent's acting on the basis of what he believes to be the weighted reasons. In robotics the concept of autonomy has a different interpretation from the philosophical one, since it focuses on the ability of independent execution of tasks. This chapter studies robot autonomy covering ontological, epistemological, and other philosophical issues. Specifically, the chapter discusses the following issues: (i) what is autonomy, (ii) what is robot autonomy, and (iii) what are the levels of robot autonomy. The other philosophical aspects of robot autonomy discussed in the chapter are: freedom of selecting goals, intrinsic ownership of goals, integration of body and control system, and relation of autonomy with intelligence and capacity.

Chapter 7: Robot Consciousness

Consciousness starts from the brain down to the genetic code and is a more complex phenomenon than intelligence. The major philosophical problems about consciousness are the existence problem (why consciousness exists?), the intelligibility problem (how do we explain and understand consciousness?), and the brain-mind correlation problem (what possible explanation could there be for the way brain processes and sensations are correlated?). In this chapter we make a tour to the basic philosophical and practical issues of human and machine consciousness. Specifically, we discuss the ontological issues of consciousness, and the ontological issues of machine-robot consciousness (purpose, plausibility, features), and we present several sets of requirements for conscious machines.

Chapter 8: Robot Conscience and Ethics

Conscience is the inner mechanism woven into our lives which guides us in the process of choosing among alternative actions considered in our conscious reasoning. This means that consciousness is a prerequisite for conscience. Both consciousness and conscience belong to the mechanisms that create the behavior of a human being. The ultimate goal of robot conscience (morality/ethics) is to create a machine that follows an ideal ethical principle or system in decisions about possible courses of action it could perform. The benefits of pursuing research on robot conscience and ethics that could lead to the creation of an ethical robot, include, among others: (i) the fact that to make ethics computable, a clear objective basis of ethics is needed to be found, (ii) the possibility that research on machine ethics may lead to the discovery of new problems with current ethical theories, and (iii) the fact that ethicists working with machine intelligence have a greater chance to achieve breakthroughs in ethics, than purely theoretical ethicists. This chapter provides an account of robot conscience and ethics, including ontological aspects, basic approaches to robot ethics (top-down, bottom-up), ethics of human-robot co-existence (robot moral status, robot rights), and ethical aspects and problems of medical, assistive, socialized, and war robots.

Chapter 9: Architectural Aspects of Cognitive Robots

The proper choice of cognitive architecture facilitates the design, implementation, and validation of mental robots. The architectural concept involves the architectural structure (subdivision of a system to subsystems, and their interconnection or interaction), and the architectural style (computational style, communication style-synchronous or asynchronous, etc.). Actually, different mental robotic applications have different needs that can be satisfied by proper architectures. In this chapter we provide an overview of principal architectural aspects, namely: (i) generic requirements and capabilities, (ii) taxonomies, (iii) the meta-cognition concept, (iv) conceptual and structural features of a number of available cognitive robots (Cog, Kismet, SOAR, ICARUS, LIDA), and (v) evaluation/validation issues. A list of 10 further cognitive-intelligent architectures with brief descriptions is also included.

Chapter 10: Additional Issues

The new way of thinking in robotics and artificial intelligence has contributed substantially in the understanding of the nature of cognition and the development of mental (cognitive, intelligent, autonomous, conscious, ethical) robots by linking more tightly sensing and acting. Two fundamental empirical questions about mental robots are: (i) whether robots and computers will be able to perform particular tasks that human performs, and (ii) if this can be done, whether robots and computers perform them in the same way that humans do. But the most important question is "whether the robot/computer's ability to perform advanced human-like tasks as a human would perform shows that the robot/computer has a mind". In this chapter we discuss a set of conceptual issues complementary to those studied in the previous chapters. These issues include: (i) consciousness theories, (ii) theory of mind, (iii) robot development, (iv) robot shared attention, (v) robot duty, legal, and ethical issues, and (vi) philosophical and legal/ethical issues of cyborgs (cybernetic organisms).

1

Background Concepts and Outline of the Book

It is not a question of either or,
robotics or humans ... The strategy we try
to employ is not an either or, but the best of both.
Sean O'Keefe

We wanted to solve robot problems and needed
some vision, action, reasoning, planning, and so on ...
Eventually robots will make everything.
Marvin Minsky

At bottom robotics is about us, it is the discipline of
emulating ourselves, of wondering how we work.
Rod Grupen

1.1 Introduction

The actual start of modern robotics is the year 1954 when Devol Jr. patented his multijoined robotic arm, and the first industrial robot named *Unimate* was put in operation by the *Unimation* (**Uni**versal Auto**mation**) company in 1961. The evolution of robots after Unimate showed an enormous expansion both in their mechanical and computational/mental abilities, as well as in the applications. Landmarks in this expansion are the Rancho Arm (1963), the Stanford Arm (1963), the mobile robot Shakey (1970), the Stanford Cart (1979), the Japanese humanoid robots WABOT-1 (1980) and WABOT-2 (1984), the Aquarobot (1989), the multi-legged robot Genghis (1989), the NASA Sojourner robotic rover (1997), the Honda humanoid ASIMO (2000), the FDA cyberknife for treating human tumors (2001), the Sony robotic dog AIBO (2003), the SHADOW dexterous hand (2008), the Toyota running humanoid robot FLAME (2010), and many anthropomorphic (humanoid) and zoomorphic social robots (sociorobots) such as HOAP, QRIO, Albert HUBO, SAYA, etc.

Today, much effort and money is given to create cognitive, intelligent, autonomous, conscious, and ethical robots for servicing the human beings in several ways, including healthcare, assistance of low mobility people, companionship, entertainment, etc.

The purpose of this chapter is:

- To introduce the reader, as a preparation for the book, to the six principal mental abilities that will be philosophically and scientifically studied in the book with reference to robots, namely: cognition, intelligence, autonomy, consciousness, conscience, and ethics.
- To provide a brief discussion of the specialized cognitive/intelligence abilities of learning and attention which are fundamental abilities for modern robots.
- To provide an outline of the book for the convenience of the reader.

1.2 Mental Capabilities of Modern Robots

Modern robots are designed to possess a variety of capabilities where perception processing and action are embodied in a recognizable *human-like* (anthropomorphic)[1] or *animal-like* (zoomorphic)[2] form in order to emulate some subset of the physical, cognitive, intelligent and social dimensions of the human (and animal) behavior and experience. Human-like and animal-like robotics attracts mechanical, electrical computer and control engineers, philosophers, psychologists, neurobiologists, linguists, and artists all over the word to contribute. Robotics with this kind of features aims to create robot beings to interact with human beings, rather than replacing them, to do more sociable work. Thus, it seems that anthropomorphic and zoomorphic robots will change the way we interact with machines. This gives rise, besides the scientific and technical questions, of a number of critical philosophical questions and controversies which have to be thoroughly addressed.

The study of anthropomorphic (and zoomorphic) robots involves two parts, namely:

---

[1]From the Greek words ἀνϑρωπος (anthropos = human) and μορφή (morphe = form/shape).

[2]From the Greek words ζῶον (zoon = animal) and μορφή.

- *Physical or body part* (mechanical structure, kinematics, dynamics, control, head, face, arms/hands, legs, wheels, wings, etc.).
- *Mental or thinking part* (cognition, intelligence, autonomy, consciousness, conscience, ethics, and related processes such as learning, attention, emotions, etc.) [1].

This book is focused on the mental part and its philosophical aspects. Figure 1.1 shows pictorially the contribution of the areas of cognition, intelligence, autonomy, consciousness, conscience, and ethics in achieving mental robots.

These components are overlapping and interrelated in complex ways which are not yet uniquely defined and described technologically, psychologically and philosophically. Thus, making robots acting mentally (like humans) is never an easy job.

A brief preparatory account of these components is as follows.

1.2.1 Cognition

Cognition refers to the full functioning of the brain at the higher level, not directly involving the details of the neurophysiological brain anatomy. It is not a distinct module of brain, or a component of the mind that deals with rational planning and reasoning, and acts on the representations acquired by the perception apparatus. It also includes subconscious, intuitive, and affective feelings and experiences of human beings. Cognition is studied within the

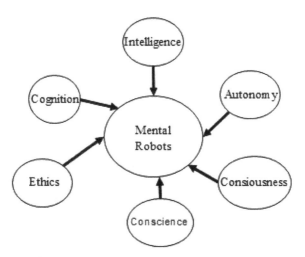

Figure 1.1 Mental robot components/capabilities.

frameworks of psychology and philosophy [2]. According to Thomas Aquinas, the psychological study of human behavior involves two paths; namely:

- *Cognitive part* (how we know the world).
- *Affective part* (how we understand the world via feelings and emotions).

In the broad sense *cognitive science* is the science that merges computer science/artificial intelligence, psychology, philosophy linguistics, and robotics (among others), and aims at understanding how intelligent beings act. The main ability of cognition is "autonomous perception" which receives and processes the external stimuli, recognizes and stores the perceived patterns, and provides the result of perception after proper matching of these patterns (Figure 1.2).

Cognition also implies our ability to understand how things might possibly be, not only now but at some future time and, based on this, to decide how to act (anticipatory action ability). Other abilities of cognition include assimilation, adaptation, action, self-reflection, and motivations that drive perceptual attention, action selection, and system development both for the short-term and long-term robust behavior. Cognitive systems engineering studies the cognition and structure both at the individual and group levels as shown in Figure 1.3.

Cognitive robotics is an important emerging field of cognitive engineering and robotics that cannot be defined in a unique, globally accepted, way. Usually a cognitive robot is defined through its required capabilities. Two major of them are [3]:

- Fluent, coupled real-world action-taking.
- Fluent real-world coupling and the ability to improve such engagements via the use of decoupled off-line reasoning.

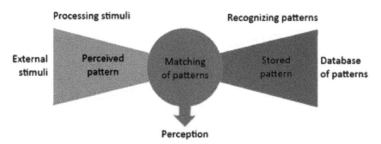

Figure 1.2 Functional structure (basic steps) of autonomous perception.

Source: http://www.daalmans.nl/en/literature/Perception,%20basic%20process.jpg

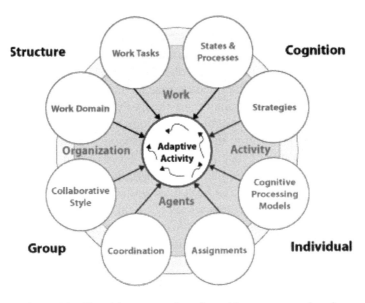

Figure 1.3   Pictorial representation of cognitive systems engineering.

Source: http://www.ppi-int.com/training/cognitive-systems-engineering-diagram.png

As we will see in Chapter 4, the philosophical aspects of cognitive science must be considered from two points of view, i.e., philosophy *in* cognitive science, and philosophy *of* cognitive science. The first deals with the philosophy of mind, philosophy of language, and philosophy of logic. The second is concerned with questions about cognitive models, explanations of cognition, correlations of causal and structural nature, computational issues of cognition, etc.

Cognitive robotics may be regarded as the engineering subfield of embodied cognitive science, and comes from the field of 'cybernetics' initiated by Norbert Wiener as the study of communication and control in living organisms, machines, and organizations [4].

1.2.2 Intelligence

Intelligence is the human cognition ability for solving problems in life and society. Individuals differ from one another in their ability to understand complicated processes, to adapt effectively to the environment, to learn from experience, and to reason under various conditions. These differences may be substantial and are not consistent since they actually vary from situation to situation or from time to time. Theories and concepts of "intelligence" attempt to organize and explain such phenomena, and, although in some areas this is done with relative success, no unified conceptualization has so far been developed that resolves all the important problems.

Human intelligence can be amplified or enhanced by practicing several well-developed tools such as mnemonics, problem-solving heuristics, creativity methods, and decision-making tools. Psychology has developed several methods for measuring or judging the level of a person's intelligence; the so-called 'psychometric intelligence tests'. On the basis of them, developmental psychologists study the way children come to think intelligently, and distinguish mentally retarded children from those with behavior problems.

With the advancement of the computer science field, many attempts were initiated to create and study computers and machines that possess some kind and level of intelligence (problem solving ability, etc.) analogous to human intelligence, which is still the subject of strong controversy [5]. Robots represent a class of machines that can be equipped with some machine intelligence which can be tested using the so-called "Turing Test". As with human intelligence, in robot and machine intelligence many philosophical questions came to be addressed, the two dominant of which are whether intelligence can be reproduced artificially or not, and what are the differences between human intelligence and machine intelligence (if it exists). Some of these questions will be discussed in Chapter 5.

1.2.3 Autonomy

Autonomy is a concept that has a multiplicity of meanings and ways in which it can be understood, interpreted and used in human's life. Four closely related fundamental meanings are the following:

- The capacity to govern oneself.
- The actual condition of governing oneself.
- The "sovereign" authority to govern oneself.
- An ideal of character.

The first meaning is understood as the basic autonomy (minimal capacity) which refers to the ability to act independently, authoritatively, and responsibly. This capacity can be capitalized either legally or philosophically. The legal right to autonomy depends largely on a person's cognitive ability to make particular conscious decisions under various circumstances on the basis of the relevant information, and evaluate their consequences.

The second meaning of autonomy mirrors the one's entitlement to certain liberal rights that determine our political status and freedoms. However, in actual life, having the capacity to govern ourselves, does not imply that we can actually do so.

The third meaning distinguishes *de jure* and *de facto* autonomy, where the former refers to the moral and legal right to self-government, and the latter to the competence and opportunities to exercise that right.

Finally, the fourth meaning (autonomy as ideal) refers to our moral autonomous agency predicated upon "autonomy virtues", with the aid of which we can correctly guide our agency and orientate public social policies concerned with fostering autonomy.

In robotics, autonomy is interpreted as independence of control. This meaning implies that autonomy characterizes the relation between the human designer and controller, and the robot. A robot's degree of autonomy is increased if the robot possesses increasing abilities of self-sufficiency, learning, development, and evolution. The two forms of robot autonomy are [6]:

- *Weak autonomy* (the robot can operate and function free from any outside intervention).
- *Strong (or full) autonomy* (the robot can make choices on its own and is authorized to activate them).

If a robot combines intelligence with autonomy, then it is said to be an *"intelligent autonomous robot"*. The issues of robot autonomy, together with some philosophical considerations will be the subject of Chapter 6.

1.2.4 Consciousness and Conscience

These two aspects of human performance have been the subject of extensive and deep studies made by psychologists and philosophers, in their attempt to explain the functions of the human brain.

- *Consciousness* refers to the issue of how is it possible to observe some of the processes that take place in our brain.
- *Conscience* refers to the issue of how is it possible to acquire and use the knowledge of what is right or wrong.

Understanding consciousness and conscience is very difficult because we cannot observe the processes that are executed in our brain, and, more critically, we cannot observe the processes running in the brain of another person. In philosophy, consciousness is interpreted as the mind or mental abilities exhibited by thoughts, feelings and volition. In psychology, consciousness has several meanings, e.g., awareness of something for what it is or the thoughts and feelings, collectively, of an individual or a society. Extending consciousness to machines and robots is not an easy issue. According to Haikonen [7], for a robot to be conscious it is required to have some kind of mind, to be self-motivated, to understand emotions and language and use it for natural communication, to be able to react in emotional way, to be self-aware, and to perceive its mental content as immaterial. Philosophers may argue that the above requirements are not sufficient for consciousness, but building such robots may help to understand better the problems of human consciousness. In [7], an engineering approach that would lead towards cognitive and conscious machines is outlined, using neuron models and associative neural networks made from the so-called "Haikonen associative neurons".

Pitrat [8], provides a comprehensive study of human consciousness and conscience and investigates whether it is possible to create "artificial beings" (robot beings, etc.) that possess some capabilities analogous to those that consciousness and conscience give to human beings. He argues that "if a system or machine (like a robot) generates a behavior similar to ours, it is possible that we are also using it". However, as he says, many artificial intelligence workers are not interested to understand how we work, but to realize systems that are as efficient as possible, with necessarily modeling of the brain.

Human beings have a warning mechanism built into them which warns them that they are malfunctioning. In other words, the resulting warning signals manifest that a human does not behave in harmony with his own values and beliefs. This mechanism is the *human conscience*. There exist two kinds of conscience [9]:

- *Conscious conscience* which all of us recognize.
- *Unconscious conscience* in which the warning signal manifests a judgment not well understood or poorly interpreted, or totally not understood.

Of course conscious conscience is what is primarily important in actual life. Unconscious conscience may exist because man is not a single system, but several interconnected systems hierarchically arranged as described by cybernetics, i.e. a multi-component feedback regulatory (homeostatic) system with two types of feedback namely informational and authoritative feedback [4].

Building a mechanism of this type into a robot we get a robot with conscience which is determined by the values and norms embedded in it. The question whether it is possible to build conscience into robots, or not, has been a diachronic issue of study in psychology, philosophy and robotics. The robot conscience topic will be discussed in Section 8.3.

1.2.5 Ethics

Ethics is a branch of philosophy that deals with the study and justification of moral beliefs which belong to human conscience. The term ethics is derived from the Greek word $\acute{\eta}\theta o\varsigma$ (ethos) meaning *moral behavior* (or *character*). Morals, from which morality comes, are social rules or inhibitions from the society. Ethics is actually the implementation of conscience in a well structured way. The development of ethics and morality throughout the history of human experience was made via religion, philosophy, society/culture, environment, and family as shown in Figure 1.4.

Robot ethics is the extension of ethics to robots which embeds ethical rules in a *top-down* or *bottom-up* (developmental) way [10]. Robot ethics involves the ethics of the people who design and use robots, and the ethics of how robots are treated by humans [11]. Thus the philosophical questions of robot ethics involve combinations of the questions of human ethics and questions that have to do with the robots. These questions are studied in Section 8.4.

Figure 1.4 Pictorial illustration of the origins of ethics.

Source: http://www.nursingworld.org/ClientResources/Images/OJIN/FIGURE.GIF

1.3 Learning and Attention

Here, we will discuss two more mental capabilities of modern robots belonging to cognition and intelligence, namely: *learning* and *attention*.

1.3.1 Learning

Learning is one of the most fundamental cognition and intelligence processes. It involves acquisition of new knowledge and behavior, and is a form of change in behavior pattern(s). Learning occurs continuously throughout a person's life. Learning theory is a set of principles developed by psychologists and educators to explain how human beings acquire skills, knowledge, and attitudes. The two basic learning theories are:

- *Behaviorism* This theory emphasizes the importance of having a particular form of behavior reinforced by a teacher to shape or control what is learned. The teacher has several ways to provide students stimuli, induce the desired behavior or response, and reinforce the behavior with proper rewards.
- *Cognitive theory* Unlike behaviorism, the cognitive theory focuses on what is going inside the learner's mind. In other words, cognitive theorists advocate that learning is not just a change in behavior, but a change in the way a learner thinks, understands, or feels.

Although psychologists and educators do not completely agree, there is a consensus that, actually, learning can be explained by a combination of behavioral and cognitive theories. The two major branches of cognitive theory are:

- *Information processing model* The student's brain has an internal structure which chooses and processes incoming information, stores and retrieves it, uses it to create behavior, and receives and processes feedback on the results.
- *Social interaction model* People learn by watching other people perform a behavior and the outcome of this behavior. This model was coined by Bandura [12], and explains the human behavior in terms of continuous reciprocal interaction between cognitive and environmental influences.

Necessary conditions for effective social learning are: *attention* (one pays attention to learn), *retention* (remembering what one paid attention to), *reproduction* (reproducing the image), and *motivation* (having a good reason to imitate).

In humans, learning is performed according to the following styles (Figure 1.5).

- *Visual* (Learn through seeing. The learner observes the body postures and the facial expression of the teacher to understand fully the content of a lesson. Here, self-learning is best via reaching, visual materials, pictures, videos, etc.).
- *Auditory* (Learn through listening. Learners follow verbal sessions and interpret the underlying meaning of speech through listening the tone of voice, pitch, speed, etc. Text reading is done aloud. Here a task can be performed after sole listening).
- *Kinesthetic-Tactile* (Learn through physical moving, doing and touching. Here, learning is best via a hands-on-approach, actively exploring the physical world).

Learning has the following characteristics:

- *It is purposeful* (People learn on the basis of what they want to do and achieve. Their goals may be short-term or for a lifetime).
- *It is the result of experience* (The instructor cannot learn for the learner, because learning is an individual process. The learner can learn only from personal experience).

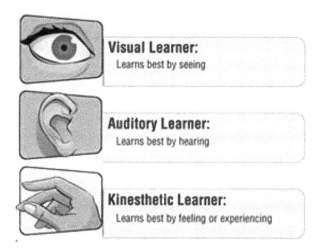

Figure 1.5 Human learning styles.

Source (Animators at Law): http://1.bp.blogspot.com/_3orqwiUwcj4/TLb5SBxk6_I/AAAAAAAAAAM/
w9CnrW5hHDk/s1600/Learning_Types.jpg

- *It is multifaceted* (Learners learn much more than expected if they fully exercise their minds and feelings. Many types of learning such as verbal, conceptual, perceptual, problem solving, emotional, situated, motion, descriptive, etc. may be combined in one learner).
- *It is an active process* (Learning is a process of changing behavior, and so it must be an active one. Humans learn better not passively, but by reacting and responding either outwardly or inwardly only intellectually and emotionally).

Robot/machine learning can be performed by using methods and processes similar to human learning [13, 14]. Robots can be taught concepts, how to acquire information, how to use their own sensors, to express emotions, and to navigate, and even how to teach themselves. Modern robots can learn by adjusting parameters, building environmental models such as maps, evolving rule sets, exploiting patterns, generating behaviors, devising new strategies, by predicting environmental strategies, and so on. A first issue to be decided is the necessary extent for the learning, and a second issue is whether learning should be performed *on-line* (while accomplishing a task in the real world) or *off-line* in a simulated environment. The most interesting robot learning takes place when the robot can create its own approach from the bottom up (i.e., without being top-down programmed for such interactions).

Robot learning can be implemented in the following ways:

- *Learning through neural networks:* The neural network learns in a supervised manner from a trainer, by adjusting its synaptic weights and thereby change the way it will respond.
- *Learning through reinforcement:* The learning takes place in an unsupervised manner using an adaptive *critic* approach, and the mappings from percepts to actions are learned inductively through trial-and-error. The critic guides the development of learning by *reinforcing* or *punishing* actions, or intended actions (Figure 1.6).
- *Evolutionary learning:* This is a kind of unsupervised learning where behavior is evolved from a randomly generated population of candidate controllers in analogy to natural evolution using some fitness function.
- *Imitative learning:* Learning by imitation is performed by showing the behavior to the robot rather than by telling it how to behave. In principle, this type of learning mjnmay employ any of the means described above namely, supervised neural networks, reinforcement, or evolution.

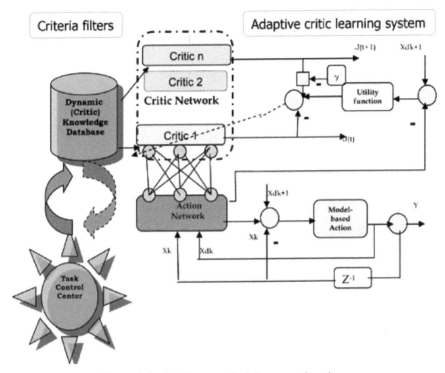

Figure 1.6 Architecture of reinforcement learning.

Source: http://spie.org/Images/Graphics/Newsroom/Imported/144/144_fig1.jpg

Imitative learning is very popular in social (socialized or socially intelligent) robots [15, 16]. Of course due to the variations in the scope and complexity of social robots the combination of the above learning types should be specialized. In addition, a social robot should have the capability to self-develop, i.e., to enrich its cognitive capacity to learn and extend its physical ability to exploit learning. This type of developmental social robot learning has been extensively studied in the "developmental psychology field" through the study of the social learning mechanisms of infants and children (now extended to adults). The available knowledge on social learning in animals is also a very useful source of inspiration for social robots, both physical and simulated.

Figure 1.7 gives an instance of child-learning by imitation.

Figures 1.8 and 1.9 give two instances of robot-learning by imitation.

Imitation learning can also work in the opposite direction, i.e., human learning from robots. This is typically used as a therapeutic aid of children

Figure 1.7   Girl learning telephone use by imitation.

Source: http://www.autism-community.com/wp-content/uploads/2011/12/Girl-Imitating-Adult.jpg

Figure 1.8   The walking humanoid social robot NAO learns dancing by imitating a human.

Source: http://hrl.informatik.uni-freiburg.de/pics/imitation_nao_freiburg_m.jpg

with autism via social robots as illustrated in Figure 1.10 [28]. Figure 1.11 shows another case of human imitation learning from a robot.

Critical questions that have to be answered for achieving such "developing" social robots include [17]:

- How a social robot can direct its own development?
- How this development is motivated and guided?
- Which *a-priori* knowledge and skill should be built in the robot?
- What bounds (if any) should be imposed?

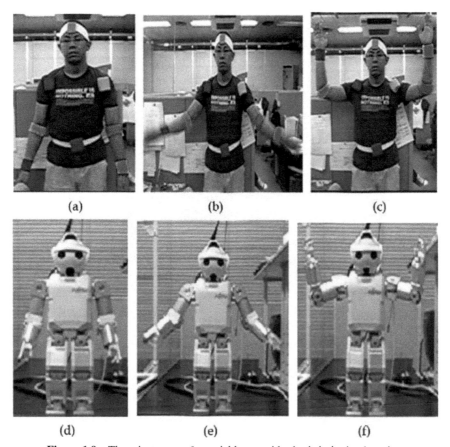

(a) (b) (c)

(d) (e) (f)

Figure 1.9 Three instances of a social humanoid robot's imitation learning.

Source: http://www.intechopen.com/source/html/6803/media/image10.jpeg

These questions may never have a definite answer, but a balance of human input, self-development and real-world interaction seems to be feasible, and actually has been realized in existing anthropomorphic humanoid robots. Many roboticists say that robot learning techniques have failed to capture the most significant property of the human brain, i.e., its capacity to not only learn, but be aware to and able to direct its own learning. Of course it is easy to argue that robot learning techniques simply provide indirect ways to tell the robot what to do, but we should not forget that good robot learning methods require interaction with a real and varying world. In general, a robot cannot learn complex skills successfully without first learning necessary simpler skills. For instance without learning how to hold a pen, the robot will not be able to

Figure 1.10   Robota social robot teaches a child through imitation learning. The second child in front observes the learning process [28].

Source: https://www.researchgate.net/publication/6368514_Building_Robota_a_mini-humanoid_robot_for _the_rehabilitation_of_children_with_autism

http://lasa.epfl.ch/publications/uploadedFiles/Ass_Tech06_inpress.pdf

Figure 1.11   The girl imitates the social robot KASPAR.

Source: http://i.dailymail.co.uk/i/pix/2011/03/09/article-1364585-0D85271C0005DC-940_468x308.jpg

learn how to write. Developmental learning differs from conventional learning in that it is not task specific, the tasks are unknown, a representation of an unknown task is generated, it is an animal-like online learning, and, generally, it is an open-ended learning [18, 19].

1.3.2 Attention

Attention is the human cognitive ability to focus selectively on a specific stimulus, sustaining that focus and shifting it at will, i.e., the ability to *"concentrate"*. It is a concept of cognitive psychology which refers to how humans actively process specific information available in the environment. Attention is important to learning, since learning is optimally efficient when a person is paying attention. The attention ability of selecting the potentially relevant parts out of a large amount of sensory data enables interactions with other human beings by sharing attention with each other. This ability is of great importance in robotics, where the computational modeling of human attention is a key issue in human-robot interaction. The ability of a robot to detect what a human partner is attending to, and act in a similar manner, enables intuitive communication which is an important desired skill for a robotic system.

Attention is triggered by two different types of influences [20]:

- Stimulus-driven (i.e., affected by bottom-up influences).
- Goal-directed (i.e., fuelled by top-down influences).

Attention is the process by which a being allocates perceptual resources to analyze an area of the surrounding world in detriment of others. This allocation can be done in two ways:

- Explicit reorientation of sensors, e.g., head reorientation in the case of visual and auditory sensors to specific part of the world (*overt attention*).
- Deployment of computational resources for processing a specific part of the sensory information stream (*covert attention*).

Overt attention is the direct cause of active perception [21]. It can be either voluntarily or involuntarily driven by automatic orientation of sensors. Usually, the overall attention starts with overt attention which is followed by covert mechanisms. Basically, the involuntary attention process is stimulus-driven but it is also modulated by goal-directed influences through *attention sets* which impose as "prioritizing measure" the task relevance.

Overt attention and active perception play a significant role for social beings, and they actually are the *precursor of intent*. Since attention is influenced by the being's goals its *overt* exhibition appears as an action and allows an external observer to get an idea of these intentions. Attention is a multi-sensory process although vision has so far been the most studied sensory type.

Actually, attention mechanisms involve the inner uncertainties of perceptual and actuation processes. Therefore, in the context of social human-robot interaction the following questions arise [22]:

- How do automatic multi-sensory attention mechanisms affect joint attention?
- How can the uncertainty inherent to attention be treated?
- How the automatic mechanisms underlying multiple sensory modalities and actions interplay in the context of joint attention?

Four fundamental prerequisites for joint attention are [23, 24]:

- *Attention detection* (The capability to track the observable attention behavior of other beings).
- *Attention manipulation* (The capability of affecting the focus of attention – **FOA** – of other beings).
- *Social coordination* (The capability to engage and regulate social interaction).
- *Intentional understanding* (The capability to view oneself and the others as intentional beings, and as such to acknowledge each other's intentions).

These requirements are connected (intrinsically and fundamentally) to automatic attention-related mechanisms. An overview of system architectures that are important in attention of robots is provided in [25], with emphasis on multimodal information fusion and humanoid robots. In all of them, knowledge concerning what needs to be attended to, and when and what to do in a meaningful sequence (in correspondence with visual feedback) is properly applied in different ways. A discussion of attention mechanisms for the imitation of goal-directed action in developmental robots is provided in [26]. A study of attention gestures in human-robot dialogues is presented in [27]. Attention gestures constitute evidence of the engagement process by which participants start maintain and end their perceived connection to one another.

1.4 Outline of the Book

The book involves 10 Chapters. Chapters 4 through 8 constitute the main body of the book, Chapters 1 through 3 provide preparatory material, Chapter 9 has a pure technical nature, and Chapter 10 contains a discussion of some additional issues which complement the material of the main body of the book. Each chapter of the main body starts by first discussing the corresponding mental ability in humans, and then extends the discussion to robots.

Chapter 1, the present chapter, gives a brief sketch of the six mental capabilities of human beings, namely cognition, intelligence, autonomy, consciousness, conscience, and ethics, which are desirable for inclusion in modern mental robots. Two other specific mental capabilities that belong to cognition and intelligence, and are basic for designing modern robots, viz., learning and attention, are also outlined. These two capacities will be discussed further in Chapter 10.

Chapter 2, *"Philosophy: Fundamental Elements"*, is the first of two chapters aiming at introducing the reader to the field of philosophy in a very elementary way, including the basic features and the subfields of philosophy, namely: metaphysics/ontology, epistemology, teleology, ethics, and aesthetics. The chapter provides the basic definitions, along with some fundamental questions in each subfield.

Chapter 3, *"Philosophy of Science and Technology"*, discusses the features of science that discriminate it from non-science, and provides a list of fundamental questions addressed in philosophy of science. These questions include the questions about whether science is philosophy or philosophy can be scientific. Then, the chapter examines the general aspects of the philosophy of technology (and engineering) including the ontological issues of artifacts and engineering design.

Chapter 4, *"Robot Cognition"*, is concerned with the fundamental philosophical aspects and questions of human and robot cognition. Starting with the ontological issues, the chapter provides the major historical landmarks of cognitive science and cognitive robotics. Some technological elements and the principal requirements for designing cognitive robots are also briefly discussed.

Chapter 5, *"Robot Intelligence"*, examines basic philosophical aspects of human and machine (robot) intelligence, and provides the major milestones of this field. The chapter summarizes the philosophical opinions of eminent scientists about the ontological and phenomenological issues of machine/robot intelligence, and reviews the Turing test and some machine intelligence tests including multi-dimensional ones.

Chapter 6, *"Robot Autonomy"*, is devoted to the autonomy issues of modern robots which include the types and the levels of autonomy. It starts with the ontological aspects of human autonomy and extends them to robots. Philosophical considerations about the freedom of selecting goals, intrinsic ownership of goals, and the relation of autonomy with intelligence and capacity are also discussed.

Chapter 7, *"Robot Consciousness"*, deals with the human consciousness issues presenting the philosophical views of several well known workers in the field. These issues are then investigated for the case of robot beings, including the purpose, the plausibility, and the features of robot consciousness. The chapter ends with the presentation of several requirements for conscious machines, as presented by a number of artificial intelligence and robotics workers.

Chapter 8, *"Robot Conscience and Ethics"*, discusses the ontological issues of human and robot conscience, and presents the basic approaches to robot ethics (top-down and bottom-up approaches), including the ethics of human and robot co-habitation (symbiosis). Then, it investigates the ethical problems arising in medical, assistive, social, and war robots.

Chapter 9, *"Architectural Aspects of Cognitive Robots"*, deals with technical aspects of cognitive robots, namely: generic requirements and capabilities of cognitive architectures, taxonomies of cognitive architectures, meta-cognition issues, and evaluation issues. The chapter includes a discussion of five architectures (at some detail) and a list of 10 further architectures (without details). An exhibition of a comparative study of 20 cognitive architectures available in the literature, is also provided.

Chapter 10, *"Additional Issues"*, presents several complementary aspects of mental robots of special interest. These include a review and comparison of the principal consciousness theories (higher-order thought, global workspace, integrated information, and biological theory), an outline of the theory of mind and developmental robotics, a discussion of robot attention issues (overt attention, covert attention, shared attention) and a look at some additional philosophical, legal, and ethical issues, including the class of cyborgs (cybernetic organisms).

In overall, the book provides a good picture of the state-of-art of the philosophical questions that arise from the creation of robots with mental capabilities similar to human ones. The strongest debate (philosophical, ethical, and technological) has been in the autonomy and ethics of robot beings, especially of the war robots. The advocators of autonomous war robots argue that these robotic weapons act more accurately and more morally than humans. These arguments are strongly rejected by most roboticists and ethicists who argue that *"autonomy of firing"* should never be given to robots. Certainly, current technology does not guarantee that war humanistic laws can be programmed to robots so as to morally surpass human soldiers in facing critical situations. Moreover, these thinkers argue that "quite likely" this will be very difficult (if not impossible) to be achieved in future. In any case, machines cannot fully replace humans in war. But, the most serious ethical concern against war lethal robots is the reasonable fear that using autonomous robotic weapons will lower the barriers of conducting wars by those countries that have available such robots. Personally, I agree with these later arguments. It is a worldwide observed fact that the increasing use of robot weapons in recent years from the stronger countries has resulted in an increase of terrorist response from the side of the weaker countries, with very serious consequences for the humanity.

1.5 Concluding Remarks

In this introductory and preparatory chapter we have provided a brief look to the six main capabilities of mental robots that are the subject of this book namely: cognition, intelligence, autonomy, consciousness, conscience, and ethics, along with an outline of the book.

The big question with mental robots that possess in some degree the above capabilities is how we, as humans, treat them. Humans have a special ethical status and responsibility. Every being on Earth that belongs to the *"homo sapiens"* is automatically part of a very special community that gives him/her rights and authority to rule over other beings. The symbiosis of human beings is governed by proper rules, compatible with Article 1 of the *Declaration of Human Rights* which says: "All people are born free and equal in dignity and rights. They are equipped with senses and conscience and should treat each other in a spirit of brotherhood". But this is not the case with animals. As a rule, animals are kept in captivity, often in very small spaces, mainly to feed them. The questions are: "Do robots have an inner life? How we establish it?" How can we differentiate between what comes from us and what comes from a mental robot (able to speak, think, and act like humans) itself? What about robot consciousness and conscience? Would be ethically correct, eventually, to stop developing robots with consciousness and conscience?" These and other similar questions concerning robot cognition, intelligence and autonomy will be addressed in the present book, both from a philosophical and practical viewpoint. Two web sources that present and discuss a number of important topics in robot philosophy are [29, 30].

References

[1] B. Scasselati, Theory of mind for a humanoid robot, Autonomous Robots, 12, 2002, 13–24.

[2] R. Sternberg, The Nature of Cognition, Cambridge, MA, U.S.A.: MIT Press, 1991.

[3] A. Clark and R. Grush, Towards cognitive robotics, Adaptive Behavior, 7(1), 1999, 5–16.

[4] N. Wiener, Cybernetics: Control and Communication in the Animal and the Machine, Cambridge, MA: MIT Press, 1998.

[5] S.G. Tzafestas (ed.), Intelligent Robotic Systems, New York, NY, U.S.A.: Marcel Dekker, 1991.

[6] J.M. Beer, A.D. Fisk and W.A. Rogers, Toward a Psychological Framework for Levels of Robot Autonomy in Human Robot Interaction, Tech. Report HFA-TR-1204, School of Psychology, Georgia Tech., 2012.

[7] P.O. Haikonen, Robot Brains: Circuits and Systems for Conscious Machines, New York, U.S.A.: Wiley, 2007.

[8] J. Pitrat, Artificial Beings: The Conscience of a Conscious Machine, Hoboken, NJ, U.S.A: John Wiley, 2009.

[9] A. Tracik, Conscience: Conscious and Unconscious, J. of Religion and Health, 75–85.

[10] W. Wallach and C. Allen, Moral Machines: Teaching Robots Right from Wrong, Oxford, U.K.: Oxford University Press, 2009.

[11] P. Lichocki, P.H. Kahn Jr. and A. Billard, A survey of the robotics ethical landscape, IEEE Robotics and Automation Magazine, 18(1), 2011, 39–50.

[12] A. Bandura, Social Learning Theory, New York, NY, U.S.A.: General Learning Press, 1977.

[13] T.M. Mitchell, Machine Learning, Cambridge, MA, U.S.A.: MIT Press, 2009.

[14] E. Alpaydin, Introduction to Machine Learning, New York, NY, U.S.A.: McGraw-Hill, 1997.

[15] C.L. Nehaniv and K. Dautenhahn, Imitation and Social Learning in Robots, Humans, and Animals: Behavioral, Social and Communicative Dimensions, Cambridge, U.K.: Cambridge University Press, 2009.

[16] C. Breazeal and B. Scasselati, Robots that imitate humans, Trends in Cognitive Science, 6, 2002, 481–487.

[17] M.L. Swinson and D.J. Bruener, Expanding frontiers of humanoid robots, IEEE Intelligent Systems and Their Applications, 15 (4), 2000, 12–17.

[18] Proceedings of Workshop on Development and Learning, Michigan University, East Lansing, MI, U.S.A., 5–7 April, 2000 (www.cse.msu.edu/de/).

[19] J. Weng, J. McClelland, A. Pentland, O. Sporns, I. Stockman, M. Sur and E. Thelen, Autonomous mental development by robots and animals, Science Magazine, 291 (5504), AAAS, 26 Jan. 2001, 599–600.

[20] M. Corbetta and G.L. Shulman, Control of goal-directed and stimulus-driven attention in the brain, Nature Review Neuroscience, 3, 2002, 201–215.

[21] J. Aloimonos, I. Weiss and A. Bandyopadhyay, Active vision, Intern. Journal of Computer Vision, 1, 1987, 333–356.

[22] J.F. Ferreira and J. Dias, Attentional mechanisms for socially interactive robots: A survey, IEEE Transactions on Autonomous Mental Development, 6(2), 2014, 110–125.

[23] F. Kaplan and V.V. Hafner, The challenges of joint attention, Interaction Studies, 7(2), 2006, 135–162.

[24] F. Kaplan and V.V. Hafner, The challenges of joint attention, Proceedings 4th Intern. Workshop on Epigenetic Robots, Genoa, Italy, 2004, 67–74.

[25] L. Paletta, E. Rome and H. Buxton, Attention architectures for machine vision and mobile robots, In: L. Itti, G. Rees and J. Tsotsos (eds.), Neurobiology of Attention, Amsterdam, The Netherlands: Elsevier, 2005, 642–648.

[26] I.D. Goga and A. Billard, Attention mechanisms for the imitation of goal directed action in developmental robots, Proceedings 6th Intern. Conference in Epigenetic Robotics (SCAI), Espoo, Finland, 2006.

[27] C. Sidner, and C. Lee, Attentional gestures in dialogues between people and robots, In: T. Nishida (ed.), Engineering Approaches to Conversational Informatics, New York, NY, U.S.A.: Wiley, 2008.

[28] A. Billard, B. Robins, J. Nadel and K. Dautenhahn, Building robota: A mini humanoid robot for the rehabilitation of children with autism, RESNA Journal of Assistive Technology, Vol. 19, No. 1, 2007, pp. 37–49.

[29] www.PhiloComp.net/home/robots.htm

[30] www.philosophy.ed.ac.uk/people/clark/pubs/#robotics

2

Philosophy: Fundamental Issues

"Εν οίδα ότι ουδέν οίδα".
"Only one thing I know, that I know nothing".
Σωκράτης – **Socrates**

"Je pense, donc je suis".
"I think, so I am".
René Descartes

"Philosophy is a science of the human being, of its representing,
thinking, and acting – it should present the human being
in all of its components".
Immanuel Kant

2.1 Introduction

Philosophy is about our thoughts, beliefs and attitudes, about ourselves and the world; essentially putting our thought and language in order. Doing philosophy is the first activity of stating as accurately, clearly, and as convincingly as possible what we believe and what we believe in. This *analytic* and *linguistic* understanding process arises from the obvious recognition that all works of philosophy are expressed and communicated in a physical form of symbol schemes in a particular modality, and interpreted using a specific language and interpretational framework. Philosophy is considered inside its processes and products.

The word *"philosophy"* comes from the Greek ancient composite word φιλοσοφία (philosophia), where φιλώ (philo = love) and σοφία (Sophia = wisdom). Thus philosophia is *"love of wisdom"*.

Western philosophy has its origin in the ancient Greek philosophy which was concerned with two principal questions, namely: *"τι εστίν"* (ti estin = what is) and *"τι το όν"* (ti to on = what is being) which refer to the essence of existence. The question *"what is"* was coined by the pre-Socrates philosopher Parmenides, born in the Greek colony Elea at about 515 BC. He explained that *"reality"*, i.e., *"what is"* is one (change is not possible), and *"existence"* is "timeless, uniform, necessary, and unchanging". According to Parmenides *"thinking"* and the *"thought that it is"* are the same because you will not find *"thinking"* apart from *"what is"*, because *"being is"* but *"not being is not"*. The fundamental aspect of this philosophy is to investigate ad understand the meaning of *"what"* in any case of asking what is the essence of something. Socrates, Plato, Aristotle and other ancient Greek philosophers were using the word *"what"* giving to it different interpretations. European philosophers like Kant, Hegel and Descartes developed their own philosophies using further different meanings of the word "what" and of the essence of "existence". Socrates (470–399 BC) was teaching and arguing but only to seek wisdom, i.e., the truth in any field. Plato defined *"philosopher"* as: "the one who feels no disgust in sampling every study, and who attacks the task of learning gladly and cannot get enough of it (i.e., the *"lover of wisdom"*).

Philosophy can be interpreted as the *"process of philosophizing"* (i.e., doing philosophy) in which linguistic symbol tokens of talking, hearing, reading, and writing (including diagrams, mathematics, etc.) are produced and received.

The purpose of the present chapter is to give an elementary introduction to the fundamental general issues of philosophy, as a preparation of reading the book.

These issues are:

- Basic features of philosophy.
- Fields of philosophy (metaphysics/ontology, epistemology, teleology, ethics, and aesthetics).

The material of the chapter includes the definition, the etymology, and some of the basic questions addressed in each field. Full studies of the above are presented in the literature (e.g., [1–18]).

2.2 Feature and Subfields of Philosophy

2.2.1 Features Philosophy

The primary features of philosophy (philosophizing) are the following [1, 2]:

- **Articulation**   Spelling out our ideas and views in words and sentences clearly and concisely, directly understandable in a language.

- **Argument** Supporting our ideas and views with reasons from other ideas, principles and observations to derive our conclusions and face objections.
- **Analysis** Understanding an idea by identifying and clarifying its various components.
- **Synthesis** Merging different ideas and views into a single unified perspective. For example, the Pythagorean concept of the *"harmony of the spheres"* synthesizes mathematics, music, physics and astronomy.

Doing philosophy means that we deal with issues, ideas and views in depth continually asking *"why"* and *"how"* to critically analyze underlying assumptions and arrive at the foundations of our complex knowledge structures. Philosophizing almost always refers to other philosophers in support of our own views, borrowing their arguments and examples, and quoting them for their views. Philosophy based on articulation and arguments helps us to analyze and evaluate our ideas, to synthesize our vision of ourselves and the world, and merge the various elements together in a single and unified way. Supporting our ideas with proper reasoning is different from insisting on the mere sound of a word (no matter how righteously). Many people use such words as true, reality, morality, freedom, love, etc., as *buzzwords*, i.e., words that make us feel good by simply saying them. But to express the beliefs these words are supposed to represent needs something more than merely saying the words; it needs us to do philosophy and try to answer the "why" and "how" questions for all aspects included. To define our terms is very difficult, and the definition can actually be given at the end of the thinking process rather than at the beginning.

2.2.2 Subfields of Philosophy

Philosophy involves several subfields (branches) bounded together in the sense that it is difficult to discuss a question in any one of them without referring to ideas and concepts from the others. In modern philosophy we encounter the following non-exhaustive subfields [3, 4]:

Metaphysics The study of reality and the nature of all things aiming at obtaining a comprehensive global view of the world (universe). A part of metaphysics is the field of *"ontology"* which deals with the study of *"being"*, i.e., what we mean when we say that something *"is"*, and attempts to list in order of priority the various kinds of entities that make up the universe (physical objects, properties, individuals, actions, events, bodies, minds, God, and so on).

Epistemology   The study of issues related to the nature of knowledge. These issues include, among others, questions like "what can we know?" How do we know anything?" and "What is truth?"

Teleology   This is the philosophy's sub-domain that asks about the aims and purposes of "what we do" and "why we exist".

Ethics   The study of *"good and bad"*, and *"right and wrong"*. Ethics is also known as *moral philosophy* (the establishment and defense of rules of morality and good life).

Aesthetics   The study of the nature of art and the concepts of beauty, pleasure and expression, including the experiences we get when we enjoy the arts or the beauties of nature and science or technology.

Logic   The study of good reasoning including questions like "What is rationality?", "Which arguments are good ones?", "Can logic be computationally automated?" etc.

History of philosophy   This is the philosophy's subdomain that looks at what famous philosophers of the past believed, and attempts to interpret their views in the framework of current thinking.

Now, for anything we care to be interested in, we have a philosophy which deals with the investigation of its fundamental assumptions, questions, methods, and goals. In other words for any X, there is a philosophy which is concerned with the metaphysical (ontological), epistemological, teleological, ethical and aesthetic issues of X, where X may be the philosophy itself.

Therefore we have the following philosophies:

- Philosophy of science
- Philosophy of computer science
- Philosophy of mathematics
- Philosophy of physics
- Philosophy of language
- Philosophy of artificial intelligence
- Philosophy of mind
- Philosophy of technology
- Philosophy of engineering
- Philosophy of design
- Philosophy of feedback control
- Philosophy of robotics
- Philosophy of religion
- Philosophy of politics
- Philosophy of sociology

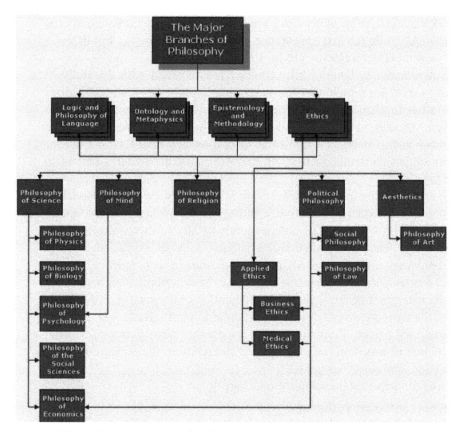

Figure 2.1 A non-exhaustive view of the major branches of philosophy.

Source: http://cas.umkc.edu/philosophy/vade-mecum/branches.gif

- Philosophy of biology
- Philosophy of psychology
- Philosophy of philosophy (meta-philosophy), etc.

A diagram showing the major branches of philosophy and their interrelation is depicted in Figure 2.1.

2.3 Ontology

The word *ontology* comes from the Greek word ουτολογία *(ontologia)* which is composed by the words *"ων"* (on) = *"being"* (to be) and "λογία/λόγος" (logia) = *"study of"* (speaking or writing about). It is noted that the etymology is Greek, but the term ontology was coined in 1606 by Jacob Lorhard and used as *ontologia*, while the first appearance of the word *ontology* in English was in the work of Gideon Harvey (1636–1702).

As mentioned in Section 2.1, ontology is concerned with the study of *"being" (existence)*, including the determination whether some categories *of being* are fundamental by asking in what sense can the entities in those categories be said, to be. This means that ontology is concerned with being in *so much* as it is being or with beings insofar as they exist (i.e., not insofar one can obtain particular facts about them or particular features belong to them) [5].

Contrary to experimental sciences that aim at investigating and modeling reality under a certain perspective, ontology is concerned with the nature and structure of things *per se*, independently of any further considerations, and even independently of their real existence. In other words, ontology in philosophy may include the study of fictitious entities (e.g., unicorn) despite the fact that they do not have actual existence, by describing their nature and structure in terms of general categories and relations. For example, in computer science the term ontology refers to a particular kind of information object or computational artifact. Computational ontologies are a way to formally model the structure of a system (i.e., the entities and relations that emerge from its observation, and serve our goals).

Some fundamental questions of ontology are:

- What can we say that exists?
- Can entities that exist be sorted, and in which categories?
- How can we model existing entities?
- What are the meanings of being?
- What does it mean for a being to be?
- Is existence a feature or a genus?
- Do fundamental beings exist, and if yes which are they?
- Are all entities objects, and how their properties relate to the objects themselves?

Parmenides view of *"existence"* is that existence is *eternal* (nothing comes from nothing) which can be considered as the prodrome (vestibule) of the physics law of *energy and matter conservation*. Therefore, our views about truth must be false or deceitful. Much of modern western philosophy has emerged from this view, e.g., *falsifiability (or refutability)*. Opposite to Parmenides, *monism* is the pluralistic view of *"Being"*, i.e., the pluralistic conception of "Being" which replaced his view that *"Being is one and unchanging"* with that of *"Becoming"* (which is an elementary ontic/existence plurality). Two particular ontic pluralities formulated in Ancient Greece are those of Anaxagoras who dealt with *"seeds"* of the various substances, and Leucippus who dealt with reality as founded on the vacuum, i.e., atoms and their intrinsic movement in it (non deterministic Atomism). This philosophical view of "Being" was later developed by Democritus in a deterministic way (4th Century BC) and further reconsidered again in a deterministic way by Epicurus. Plato has argued about the discrimination between *true reality* and *illusion* in justifying that what is real are eternal and unchanging "Forms of Ideas". Things of these Forms experienced by sensation are at best, mere copies, and real only as far as they copy such Forms. Aristotle recognizes four different dimensions of ontology, namely:

- According to several categories (or ways) of discriminating a being as such.
- According to its truth or falsity.
- According to its existence in and of itself or simply "coming along" accidentally.
- According to its potency, energy (movement), or finished presence.

Other typical ontological dichotomies are the following:

- Universals and particulars.
- Abstract and concrete entities.
- Determinism and indeterminism.
- Beliefs or realities.
- Essential (or at least relevant) and nonessential.

According to the level of abstraction, ontologies are classified as:

- *Upper ontologies* that are used for the formulation of an ontology (or meta-ontology).
- *Domain ontologies* that refer to a particular area of interest.
- *Interface ontologies* that refer to the juncture of two domains.
- *Process ontologies* that include system entities and variables (inputs, outputs, constraints, models, architectures, controllers, etc.).

2.4 Epistemology

Epistemology is about knowledge; its nature, adequacy and inadequacy. The word espistemology comes from the Greek words *"ἐπιστήμη"* (episteme = knowledge, understanding, science) and "λόγος" (logos = study of). Hence, epistemology is also called *"theory of knowledge"*. It is concerned with finding answers to questions of what knowledge is and how it can be acquired, and how it relates to concepts like truth, belief, and justification.

Traditionally, epistemology involves two approaches [6–8]:

- *Rationalism* according to which knowledge is acquired (gained) via reasoning.
- *Empiricism* according to which knowledge is acquired through sensory observation (measurements).

Generally, most philosophers agree that both these approaches to knowledge are required, and that to a certain extent they complement and correct each other.

2.4.1 Rationalism

Rationalists focus on what it is called *"necessary truth"*, meaning that certain things are necessarily true, always, and universally. Since "necessary truth" is what we know that must be true *before* we even start looking at the world and the senses reveal to us. It is also called *a priori truth* from the Latin *a priori* (before hand). The most basic form of necessary truth is the *"self-evident truth"* which means that there is no need even to think about it, since it has to be true. Some examples of self-evident truths are the following:

- Everything exists.
- Something either is or is not.
- One cannot be simultaneously (at once) in two different places.

Very often what is self-evident to one person is not self-evident to another. Some people consider as self-evident that *"the universe had to have a beginning"*, whereas other people believe that universe has always been.

To reason (infer) about more complex knowledge we have to add the so-called, *analytic truth* (derived by *deduction*), which is what we usually think of when we think of thinking. Using the rules of logic we can find, what truths follow from other truths. The basic form of deduction syllogism is a pattern coined by Aristotle *(modus ponens)* which continues to be the foundation of logic up to the present days.

The *modus ponens* rule is the following:

IF X, **THEN** Y
  X
THEREFORE Y

For example,
IF you are human, **THEN** you are mortal
You are human
THEREFORE you are mortal

Another logic reasoning rule is the *modus tollens* rule:
IF X, **THEN** Y
Not Y
THEREFORE not X

For example,
IF you are human, **THEN** you are mortal
You are not mortal
THEREFORE you are not human

It is remarked that the *denial of the antecedent* in the *modus ponens* rule does not always imply a negative consequent.

For example, the statement:

"**IF** you are human, **THEN** you are mortal
You are not human
THEREFORE you are not mortal"
is not true since animals are not humans but they are mortal.

Similarly, the *affirmation of consequent* in the *modus tollens* rule does not necessarily lead to a valid result. For example, the statement:

"**IF** you are human, **THEN** you are mortal
You are mortal
THEREFORE you are human"
is not valid since animals are mortal but they are not humans.

2.4.2 Empiricism

The moral *empiricism* is derived from the Greek word $\varepsilon\mu\pi\varepsilon\iota\rho\iota\alpha$ (empiria = experience) or $\varepsilon\mu\pi\varepsilon\iota\rho\iota\kappa\dot{o}\varsigma$ (empiricos = experiential), and is generally a *theory of knowledge* focusing on the role of experience, predominantly experience gained by perceptual (sensory) observations. In other words empiricism is based on the so-called "empirical" (or synthetic) truth which we derive from our sensory experience of the world. Empiricism is closely related to science but is not identical with it, although many of its advocators were famous scientists right from the beginning (e.g., Galileo).

The simplest form of empirical truth is that based on direct observation. To build a more general and complex body of knowledge from direct observations we use inductive reasoning which produces *indirect empirical knowledge*. We get observations and measurements and appropriately generalize them to cover more ground than we could cover directly. The questions here are "how many observations do we need to make before we can comfortably (and reasonably) generalize", and "how many exceptions to the desired result can be explained as some kind of methodological error before it becomes too large". These questions are deeply addressed by statistics.

Like rationalism, there are several variants of empiricism, most of which are forms of epistemological *realism* which accepts that what the senses show us is the reality, the truth. Realism is distinguished in:

- *Direct realism* which says that what we see is what we get and that the senses portray the world accurately.
- *Representative (critical) realism* which says that what we see is sensations, i.e., the images of the things in the world, not the things directly. Critical realists show many instances in which our eyes (and other senses) deceive us.

Direct realists argue that the so-called *illusions* are actually resulting from insufficient observations and information. Critical realism states that there are two steps for experiencing the world, namely:

- The thing itself and the light, sounds, etc. it gives off.
- The mental processing which is performed after the light hits our retinas, or the sound hits our eardrums.

Direct realism states that only the first step is sufficient.

Other tendencies in epistemology include the *idealism* which says that knowledge is primarily acquired in the form of concepts not derived from experience, and *constructivism* which says that all knowledge is a compilation of human-made constructions, not the mere discovery of an objective truth. Constructivism provides novel interpretations of knowledge and truth that constitute a new paradigm, based on inter-subjectivity instead of the traditional objectivity, and on viability instead of truth [17, 18].

Rationalism and empiricism have over the years produced joint or cooperative results, and have not been developed antagonistically. Actually, one can say that *science* is a well balanced mixture of them. Famous rationalists include Descartes, Leibniz and Spinoza, and two eminent empiricists are Hume and Locke.

Modern developments of epistemology include the *virtue epistemology* which focuses on the characteristics of the knower rather than individual beliefs or collections of beliefs, and the *epistemic relativism* which says that there is no set of rules for belief acquisition that are appropriate for all people and all situations.

2.5 Teleology

Teleology is the study of the evidences of design or purpose in nature, and was originated by the belief that certain phenomena are best explained in terms of *"purpose"* rather than *"cause"*. It is the philosophical doctrine that final causes exist in nature in the sense that nature inherently tends towards definite ends, analogous to the purpose encountered in human actions [14]. The word teleology comes from the Greek word $\tau\varepsilon\lambda\varepsilon o\lambda o\gamma\acute{\iota}\alpha$ (teleologia) composed by the word $\tau\acute{\varepsilon}\lambda o\varsigma$ (telos = end, purpose) and $\lambda\acute{o}\gamma o\varsigma$ (logos = study of). Teleology was founded by Plato, Aristotle, Immanuel Kant (as regulative principle), and was fundamental to the speculative philosophy of Hegel. According to Francis Bacon, in science, explanations based on teleology should be deliberately avoided, because no matter if they are valid or not valid they are beyond the human perception ability.

Plato asserts that true explanations for any given physical phenomenon must be teleological, and while the materials composing a body are necessary for its action or movement in some manner, they cannot be the sufficient cause for this movement or action. Aristotle asserts that Democritus view that all things can be reduced to mere necessity is wrong, since this view neglects the purpose, order, and "final cause" that brings about these necessary conditions. These views of Plato and Aristotle are opposite to those of Democritus and Lucretius who were supporters of metaphysical naturalism or accidentalism, according to which *"what happens to exist is the cause of its use"* or, in other words "nothing in the body is made in order that we may use it". Spinoza believed that human action is often *"goal-directed"*. He started the "Treatise on Emendation of the Intellect" with the remark that most human action is directed toward "wealth, honor or sensual pleasure".

From a systems theory point of view *goal-directed (purposeful)* active behavior is distinguished in two categories:

- Feedback (or teleological) behavior.
- Non-feedback (or non-teleological) behavior.

Feedback can be *positive* (in which case the fraction of the system output that returns to the system has the same sign as the original input signal) or *negative* (in which case the output fraction that reenters the system has opposed sign). In the negative feedback case the system is controlled by the margin of the error at which the system stands at a given time with reference to a relatively specific goal. The signals from the goal are used to restrict system outputs that would otherwise go beyond the goal. All purposeful systems (behaviors) may be considered to require negative feedback, i.e., for the goal to be achieved, some signals from the goal are necessary at some time to direct the behavior. In *non-feedback systems* there are no signals from the goal that modify the system's activity in the course of the behavior, i.e., the system is merely *input driven by causality*. The behavior of most technological systems and living animals involves a continuous negative feedback from the goal that modifies and guides their behavior. From the above discussion it follows that *"teleological behavior"* is synonymous with *"behavior controlled by negative feedback"* which leads to a gain in precision by a sufficient restricted connotation. The only thing that teleology shares with causality is the *time axis*. But causality implies a one-way relatively irreversible functional relationship, whereas teleology is concerned with behavior, not with functional relationships. Teleology is popular in business where businessmen think it terms of purposeful action as, for example, in management by objectives. In medicine, teleology provides a basis for professional ethics of medicine. Doctors are generally concerned with the outcome (telos) of a given therapy protocol.

In control, negative feedback is used to stabilize an otherwise unstable system or regulate a system such that its output (product, outcome) satisfies desired specifications. Positive feedback is purposeful used in both biological organisms and technological systems to get sustainable oscillatory behavior.

2.6 Ethics

2.6.1 General Issues

Ethics is the branch of philosophy which examines what is *right* and what is *wrong*. Philosophers regard ethics as *moral philosophy*, and morals as societal beliefs [9–12]. Thus it may happen that some society's morals are not ethical because they merely represent the belief of the majority. However, there are philosophers who argue that ethics is by nature relativistic in the sense that what is right is determined by what the majority believe. For example,

Aristotle's view of ethics was that "ethical rules should always be seen in the light of traditions and the accepted opinions of the community". Some psychologists (e.g., Laurence Kohlberg) argue that moral behavior is derived by moral reasoning which is based on the principles and methods that one uses in his/her judgment. Other psychologists regard the ethical behavior as the humanistic psychology movement. For example, to determine what is right or wrong one may start from "self-actualization" which is one's highest need and fulfills his/her potential.

In analytic philosophy, ethics is distinguished in the following levels:

- Meta-ethics
- Normative ethics
- Applied ethics

Meta-ethics examines the nature of morality, in general, and what justifies moral judgments. Basic questions studied in meta-ethics are:

- Are ethical demands *true-apt*, i.e., capable of being true or not true, or are they, for example emotional claims?
- If they are *true-apt*, are they ever true, and if so what is the nature of the facts they represent?
- If there are moral trues what makes them true, and they are absolutely true or always relative to some individual or society or culture?
- If there are moral trues, one way to find what makes them true is by using a *value system*, and the question here is if there is a value that can be discovered. Socrates' and Plato's reply is *yes*. They both believe that *goodness* exists absolutely, but they do not agree on what is good. Moral antirealist philosophers argue that there are no ethical truths. For example, the modern empiricist *Humes* has the position that moral expressions are expressions of emotion, or sediment feeling.

Normative ethics attempts to find what is for an action to be morally acceptable, i.e., it attempts to provide a system of principles, rules, and procedures for determining what (morally speaking) a person should do and should not do. Normative ethics is distinguished from meta-ethics because it investigates standards for the rightness or wrongness of actions, whereas meta-ethics examines the meaning of moral language and the metaphysics of moral facts. Overall, normative ethics is concerned about the issue of how we ought to live and act, and the requirements for a human to live well. It also involves the theories of *social justice* which examine how a society must be structured and how the social goods of freedom and power should be distributed in a society.

Applied ethics is the branch of ethics that investigates the application of ethics theories in actual life. It attempts to illuminate the possibility of disagreement about the way ethics theories and principles should be applied. Some specific areas of applied ethics are [9, 10]:

- Medical ethics
- Welfare ethics
- Legal ethics (justice)
- Business ethics
- Computer ethics
- Automation ethics
- Robot ethics
- Information ethics

2.6.2 Ethics Theories

Key theories of ethics are:

- Virtue theory *(Aristotle)*
- Deontological theory *(Kant)*
- Utilitarian theory *(Mill)*
- Justice as fairness theory *(Rawls)*
- Value-based theory
- Case-based theory
- Social contract theory*(Hobbes)*

A brief discussion of them follows.

Virtue theory   This theory (also called character-ethics theory) is based on the concept of *virtue* which is defined to be a character a human being needs to flourish or live well. *Virtue* comes from the Latin word *virtus,* which means *excellence of a person* (the Greek word $\alpha\rho\varepsilon\tau\acute{\eta}$: areti). Virtue theory tells that "an action is right if it is what a virtuous agent would do in the situation at hand". Thus the virtue theory is actually concerned with building good character (personality) by creating traits and habits toward acting with justice, prudence, courage, temperance, compassion, wisdom, and fortitude [12]. Aristotle's two principal virtues are $\sigma o\varphi\acute{\iota}\alpha$ (Sophia = wisdom) and $\varphi\rho\acute{o}\nu\eta\sigma\iota\varsigma$ (phronesis = practical wisdom). Plato's cardinal virtues are wisdom, courage, temperance, and justice.

Deontological theory   This theory gives emphasis to the principles upon which the actions are based, rather than the actions' results. Therefore, to act rightly one must be motivated by proper universal deontological principles that treat everyone with respect (*"respect for persons' theory"*) [13]. The word deontology is derived from the Greek word $\delta \varepsilon o \nu \tau o \lambda o \gamma \acute{\iota} \alpha$ (deontologia) which is composed by two words: *"$\delta \acute{\varepsilon} o \nu$"* (deon = what is needed) and $\lambda \acute{o} \gamma o \varsigma / \lambda \acute{\varepsilon} \gamma \omega$ (logos/lego = study of). Kant's ethics is based on the concept of *"categorical imperative"*. His model of practical reasoning is based on the answer to the question: "how do I determine what is rational?" Here rationality means *"do what reason requires"*, i.e., without inconsistent or self-contradictory policies. The reason why Kant does not base ethics on consequences of actions but to duties, is that, in spite of our best efforts, we cannot control the future.

Utilitarian theory   This theory belongs to the consequentialist theories of ethics that are *teleological* and aim at some goal or state, and evaluate morality of actions on the basis of progress toward the goal. More specifically, utilitarianism measures morality on the basis of the maximization of net expected utility for everyone affected by a decision or action. The fundamental principle of utilitarianism is: "Actions are moral to the extent that they are oriented towards promoting the best long-term interests *(greatest good)* for everyone concerned". According to Mill the good is defined in terms of well-being (pleasure or happiness) which is the Aristotelian $\varepsilon \upsilon \delta \alpha \iota \mu o \nu \acute{\iota} \alpha$ (eudaimonia = happiness). The utility principle tries to bridge the gap between *empirical* facts and normative conclusions using a pure cost/benefit analysis, where each one person should be counted as only one and not more than one.

Justice as fairness theory   This theory combines the Kantian and utilitarian philosophies for the evaluation of social and political bodies. The *justice as fairness* theory is based on the following principle: "General primary goods, viz. liberty and opportunity, income and wealth, and the bases of self-respect are to be distributed equally, unless an unequal distribution of any or all of these goods is to be to the advantage of the least favored". This principle involves two parts: *the liberty principle* (each human has an equal right to the widest basic liberty compatible with the liberty of others), and the *difference principle* (economic and social inequalities must be regulated such as they are reasonably expected to be to everyone's benefit attached to positions and offices to all).

Value-based theory   This theory uses some *value system* which consists of the ordering and prioritization of ethical and ideological values that an individual or community holds. Value is *what* a person wants to do. It is not a deontological action but a *want-to-do* action. Two individuals or communities may have a set of common values but they may not have the same prioritization of them. Values are distinguished in:

- *Ethical values* (which specify what is right or wrong, moral or immoral, and define what is permitted or prohibited in the society that holds these values).
- *Ideological values* (which refer to more general or wider areas of religion, political, social, and economic morals).

A value system must be consistent, but in real-life this may not be so.

Case-based theory   This is a modern ethics theory that attempts to overcome the apparently impossible divide between deontology and utilitarianism. It is also known as *casuistry theory*. Casuists start with immediate facts of a particular case itself, and then examine what are morally significant features (both theoretical and practical). Casuistry finds extensive application in juridical and ethical considerations of law ethics. For example, lying is always not permissible if we follow the deontological principle. However, in casuistry one may conclude that a person is wrong to lie in formal testimony under oath, but lying is the best action if the lie saves life.

Social contract theory   This theory is based on the view that the ethical duty of persons is dependent upon a contract or agreement among them to form the society in which they belong. Although not explicitly stated, the moral argument of Socrates to convince Crito why he must remain in prison and accept the death penalty was actually a kind of social contract argument. The social contract theory was developed as an independent ethics theory by Thomas Hobbes (1588–1679), John Locke (1632–1704), and Jean-Jacques Rousseau (1712–1778). It has enormously influenced the moral, social, and political life of modern West civilization. In the twentieth century moral and political theory regained philosophical momentum after John Rawls has proposed a Kantian version of social contract theory (the Justice as Fairness theory) [19]. Social contract theory has received much criticism, particularly by feminists and race-conscious philosophers [20, 21].

A pictorial representation of four main ethical theories (utilitarian, Kantilianism, social contract, character/virtue ethics) is given in Figure 2.2.

The fundamental ethical selection and decision-making stages for resolving actual ethical dilemmas in society and technology are depicted in Figure 2.3.

2.7 Aesthetics

Aesthetics (also spelled *esthetics*) is the philosophical study of beauty and taste, and is closely related to the philosophy of art which is concerned with the nature of art. The term aesthetics comes from the Greek word

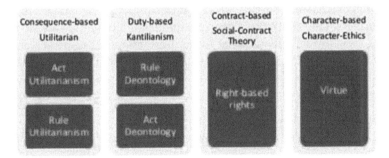

Figure 2.2 A summary of four basic ethical theories.

Source: http://image.slidesharecdn.com/ethicaltheories-141019232233-conversion-gate02/95/ethicaltheo
ries-2-638.jpg?cb=1413761040

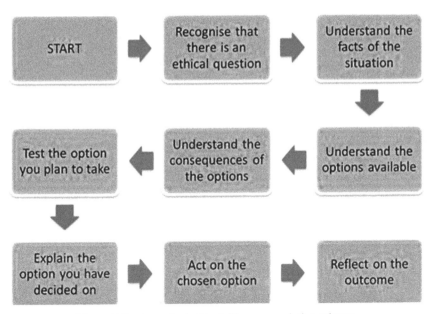

Figure 2.3 A practical ethical dilemma resolution scheme.

Source: www.ia.govt.nz/ethics-toolkit/solving-problems.asp

αισθητική (aisthetike = sense perception). Aesthetics has traditionally been part of philosophical fields like epistemology or ethics. It started to come on its own and become a more distinct branch of philosophy by Kant who viewed aesthetics as unitary and self-sufficient form of human experience [16]. The concept *aesthetics* cannot be easily broken into simpler ideas, and so it is difficult to fully explain it. Speaking for something that provides an aesthetic experience we usually talk about some form of art, although aesthetics and art are not equivalent fields. Questions about the nature of beauty and taste include:

- What is beautiful?
- Why do we find and taste certain things beautiful?
- Do we enjoy sights and sounds (music, etc.) because they are beautiful, or are they beautiful because we enjoy them?
- Does the enjoyment of beauty involve a particular sort of experience? If this is the case how should we define it and what psychological capabilities does it presuppose?
- How exactly aesthetic judgments are unified across context and art form? We call a person, a house, a robot, a car, a mathematical proof all beautiful. Are they all beautiful in the same way?
- Is beauty in the natural world quite different from artificially created beauty?
- Does there exists some underlying unity to aesthetic judgment, and are there some ways to articulate the similarities of a beautiful sunset, beautiful house, beautiful mathematical proof, etc.?
- What counts as art? How best can we define the term "art" as a subject of contention?
- What should we judge when we judge art?
- What is the value of aesthetics and art?

Aesthetics leads to a variety of issues involving morality, politics and so on. Many philosophers of aesthetics have argued that an important component of the aesthetic experience is the wish for political action. Thus, "good" art is that which gets us to try to improve society. Historically, ancient aesthetics was largely, but not entirely, based on the six great ancient civilizations, viz., Egypt, Mesopotamia, Greece, Rome, India, and China. Each of these cultures developed a unique and characteristic style in aesthetics and art. Greece had the greater influence on the development of Western aesthetics. Plato argued that beautiful objects incorporate promotion, harmony, and unity among their parts. Aristotle found that the universal elements of beauty were order, symmetry, and definiteness. Western medieval aesthetics was highly religious in focus. Modern Western aesthetics (late 17th to early 20th century) showed a slow evolution to what is called *modernism*.

Aesthetics is, and must be, applied to all societal activities, e.g.:

- Visual arts
- Music
- Literature
- Mathematics
- Information technology
- Cognitive science
- Technology
- Robotics
- Automation
- Industrial production

2.8 Concluding Remarks

In this chapter we presented the fundamental issues of philosophy. Western philosophy has its origin in the work of ancient Greek philosophers *(lovers of wisdom)* Socrates, Plato, Aristotle and others who have developed the fundamental concepts of philosophizing (articulation, argument, analysis, and synthesis). Doing philosophy means the deep study of issues, ideas and views continually asking "why" and "how" in order to analyze underlying assumptions and arrive at the foundations of human knowledge structures. Specifically, the chapter has reviewed the subfields of philosophy (metaphysics, epistemology, teleology, logic, ethics, and aesthetics), focusing on the definition, the etymology, and the basic questions addressed in each subfield.

A usual understanding of philosophy is that of *"questioning search"* and thus *"pursuit knowledge"*. This search is not an end product of such a search as a bulk of knowledge or information. Since humans are finite and mortal beings, this search pursuit is an endless process as the Athenian statesman, law maker and Poet *Solon* (638–558 BC) pointed out in his statement "Γηράσκω αεί διδασκόμενος" (pronounced "girasko aei didaskomenos") which means: "I grow old learning something new everyday". In modern Greek, this statement is expressed as "όσο ζω μαθαίνω (oso zo matheno)", i.e., "as long as I live, I learn".

Given the fact that human's brain ability is divided in *"left brain"* and *"right brain"* ability, many thinkers claimed that genuine persons can rarely use both sides of their brain at once. But many others argue, more convincingly, that when philosophizing both sides of the brain are waked up just what philosophy requires. As it is the case of any search or exploration, in doing philosophy we do not know the end of its process. However, the availability of a kind of *"pre-knowledge"* or a kind of *"anticipatory feeling"* of what the final outcome might be, would facilitate considerably the searching process itself.

References

[1] R.C. Solomon and K.M. Higins, The Big Questions: A Short Introduction to Philosophy, Belmont, CA, U.S.A.: Wadsworth Cengage Learning, 2010.

[2] W. Buckingham and D. Burnham, The Philosophy: Big Ideas Simply Explained, London/New York: D.K. Publishing, 2011.

[3] M.K. Munitz, Contemporary Analytic Philosophy, Upper Saddle River, N.J., U.S.A.: Prentice Hall, 1981.

[4] B. Stroud, Meaning Understanding, and Practice: Philosophical Essays, Oxford, U.K.: Oxford University Press, 2000.

[5] N. Guarino, D. Oberle and S. Staab, What is an Ontology? Dordrecht/ Berlin: Springer, 2009.

[6] P.D. Klein, Epistemology, In: E. Craig (ed.) Routledge Encyclopedia of Plilosophy, London: Routledge, (1998, 2005). Retrieved May 14, 2014, from http://www.rep.routledge.com/article/P059

[7] L. BonJour, Epistemology: Classic Problems and Contemporary Responses, Lanham, MD, U.S.A.: Rowman and Little field, 2002.

[8] J. Dancy, An Introduction to Contemporary Epistemology, New York, U.S.A.: John Wiley, 1992.

[9] R. Shafer-Landau, The Fundamentals of Ethics, Oxford, UK: Oxford University Press, 2011.

[10] P. Singer, Applied Ethics, Oxford, UK: Oxford University Press, 1998.

[11] J. Bennetl, A Study of Spinoza's Ethics, Indianapolis, U.S.A.: Hackett, 1984.

[12] R. Hursthouse, On Virtue Ethics, Oxford, U.K.: Oxford University Press, 2002.

[13] S. Darwall (ed.), Deontology, Oxford, U.K.: Blackwell, 2002.

[14] A. Woodfield, Technology: Creating Success, Cambridge, U.K.: Cambridge University Press, 1976.

[15] O. Hanfling, Philosophical Aesthetics, New York, U.S.A.: John Wiley, 1992.

[16] Immanuel Kant: Aesthetics, Internet Encyclopedia of Philosophy Retrieved May 18, 2014 from www.iep.utm/edu/Kantaesth

[17] J.D. Raskin and K. Bridges (eds.), Studies in Meaning: Exploring Constructivist Psychology, New York, NY, U.S.A.: Pace University Press, 2002, 1–25.

[18] J.D. Raskin and K. Bridges (eds.), Studies in Meaning 2: Bridging the Personal and Social in Constructivist Psychology, New York, NY, U.S.A.: Pace University Press, 2004.

[19] J. Rawls, A Theory of Justice, Cambridge, MA, U.S.A.: Harvard University Press, 1971.

[20] C. Mills, The Racial Contract, Ithaca, NY, U.S.A.: Cornell University Press, 1977.

[21] C. Pateman, The Sexual Contract, Redwood City, CA, U.S.A.: Stanford University Press, 1988.

3

Philosophy of Science and Technology

Science is what you know, philosophy
is what you don't know.
Betrand Russell

We still do not know one thousand of one percent
of what nature has revealed to us.
Albert Einstein

All meaningful knowledge is for the sake of action,
and all meaningful action is for the sake of friendship.
John Macmurray

3.1 Introduction

A straight forward way to see *"what is science"* is to enumerate the subjects that are known to belong to science (physics, chemistry, mathematics, meteorology, biology, etc.) and those that do not belong to it (arts, music, philosophy, theology, etc.). However such an answer is not a true answer which can only be obtained by identifying what are the things that scientific areas are sharing, i.e., what makes something to be characterized as science. Undoubtedly science is attempting to understand, explain, and predict the world around us, but this also holds for religion, astrology, history (understanding and explaining, the past), or fortune telling that tries to predict the future. It is widely accepted that the main feature distinguishing science from non-science is the type of methods that scientists use to study the world which are dominantly based on experiment (e.g., physics) and observations (e.g., astronomy).

Philosophy of science tries to understand what are experimental methods, observations and scientific theories or models, and how they have enabled scientists to reveal so many secrets of the natural world. Philosophical considerations and questions seem sometimes far away from practical concerns. But this is not so since philosophical questions about the scientific methods, the truths and operations of natural laws and the reality can be turned out to have an importance extended to real life and society, outside philosophy.

Philosophy of technology is much more difficult to be defined than philosophy of science. It is actually a broad field crossing many other fields of human activity. Many thinkers have argued that technology is *"applied science"* but without convincingly justifying the differences between science and technology. Their argument is that technology involves two parts, viz., *"substantial* (scientific) *theories"* that specify the action made by technology, and *"operative theories"* that deal with the action itself and, although they are not preceded by scientific theories, they are the product of applied science research.

But defining technology as applied science is misleading since in our days the concept of science has been tremendously enlarged to incorporate virtually all human learning. Moreover, many technological developments have not emerged as a direct application of science. Of course, it is true that modern technology is pursued primarily by people with a scientific background and within the framework of modern science, but many technological inventions have occurred by chance or by trial-and-error, not via direct application of scientific theory to get a specific artifact. Modern technology has shown many accidental inventions. As it will be described in the sequel, today it is generally accepted that technology is concerned *"what is to be"* or *"should be"* (not with *"what is"*), i.e., technology aims to change the world at will. In this respect there is a consensus on defining technology via the concept of *technological system*, which includes not only the machines but also the people involved (designers, engineers, operators, decision makers, users, etc.). Philosophy of technology is focused on technology itself and attempts to study the nature of artifacts, and understand the process of designing and creating artifacts, and man-made processes and systems.

The purpose of this chapter is:

- To discuss the question of what is science and provide a list of fundamental questions addressed in philosophy of science.
- To examine the distinction of science from non-science, and the falsiability principle.
- To discuss the philosophical principles of rationalism and logic reasoning (deductive, inductive, abductive, non-monotonic), and the principle of empiricism versus idealism.

- To present the logical positivism and paradigm model theories of confirmation.
- To examine the questions whether science is philosophy or philosophy can be scientific.
- To study the general aspects of the philosophy of technology and its fundamental questions (what is technology, what is the ontological status of artifacts, what is engineering, what is design, what is philosophy of design, etc.).
- To give a hierarchical meta-theoretical taxonomy that facilitates the critical analysis of design.

Comprehensive studies and answers to the above and other questions for deep reading in philosophy of science and philosophy of technology can be found in the literature [1–20].

3.2 Philosophy of Science

3.2.1 What Is Science?

Very broadly, *science* is any system of knowledge that aims to systematically study the physical world and its phenomena. The word "systematically" means the use of systematic or particular methods, such as observation and experimentation to get a deeper understanding of our world. Generally, a science is concerned with the pursuit of knowledge that covers the truths, and operations of fundamental world laws. Science can be and has been defined in several ways which, however, appear to involve a number of common stages for the process of scientific inquiry.

The principal stages of this process are the following, in bottom-up hierarchical order:

- Make scientific observations
- Formulate scientific questions
- Design scientific experiments
- Collect scientific data and information
- Develop scientific interpretations
- Evaluate scientific hypotheses
- Study scientific conclusions and implications
- Evaluate different points of view
- Make and test deductive predictions, testing the premises or the hypothesis via experimentation

The term *"science"* comes from the Latin word *"scire"* which means *"to know"*. The word *"scientist"* is a relatively modern word coined by William Whewell (1794–1866) in Britain. Prior to this, scholars who studied science were called *"natural philosophers"*, and include philosophers who provided accounts of the purposes and methods of the sciences. For example, Aristotle was the first great biologist and deeply studied *logic* (Aristotelian logic), René Descartes (1596–1650) formulated *analytic geometry* (Cartesian geometry), and Gottfried Wilheim Leibnitz (1646–1716) is the founder of *calculus*. Francis Bacon in his work *"Novum Organum"* (1620), a reference to Aristotle's Organon, presented a new system of logic improving the old philosophical process of syllogism. This system of logic relied on experimental histories to eliminate alternative theories.

In general, these and other thinkers studied the world around them first through *ideas* (philosophy), then as *observation* and *discovery,* and then as

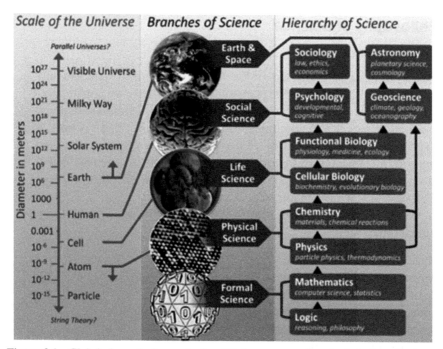

Figure 3.1 Global view of science: Scale of Universe mapped to the branches of science and the hierarchy of science.

Source: https://upload.wikimedia.org/wikipedia/commons/thumb/7/75/The_Scientific_Universe.png/500p x-The_Scientific_Universe.png

application (what is now called technology). It was believed that science (scientist, scientific process) is neutral and objective, free of personal biases, outlook, opinions, and worldview. But history of science has shown that this is not the case. As noted above, the scientific method is really only one of several ways of confirming scientific knowledge. The study of this variety of methods and views of science is the subject of philosophy of science which is concerned with the foundations, methods, implications, and, in general, all the elements of scientific inquiry from a philosophical perspective. For the convenience of the reader we give in Figure 3.1 the full mapping of the scale of universe to the branches of science and hierarchy of science namely: *Scale of the Universe → Branches of science → Hierarchy of science.*

3.2.2 Philosophy of Science: Fundamental Questions

The principal questions of philosophy of science are concerned with the purpose of science, the validity/reliability of scientific theories, and, above all, with the question of what counts as science.

Some more specific questions investigated by philosophy of science are:

- What is law of nature?
- How can science and non-science be distinguished?
- What type of data is needed to be used for distinguishing real causes and accidental regularities?
- What kinds of evidence and how much evidence are needed before the acceptance of hypotheses?
- What makes scientists continue to rely on models and theories which are today known to be at least partially incomplete and inaccurate (e.g., Newton mechanics)?
- Can one scientific discipline be reduced to the terms of another?
- Can science reveal the truth about unobservable things?
- Can scientific reasoning be justified at all?
- Is the human nature objective or is inevitably shaped by personal values and social relations?
- Is science philosophy?
- Could (should) philosophy be scientific?

Clearly, the above questions can be answered by invoking ontology, epistemology and logic concepts and issues. In the following, a short account of the answers to some of the above questions, given over the history of science by eminent philosophers, is presented.

3.2.3 Distinguishing between Science and Non-Science, and Falsiability

The problem of reliably distinguishing science from non-science is called the *"demarcation problem"*. Most modern philosophers of science largely agree that no single, simple criterion exists that can be used to demarcate the boundaries of science, while other philosophers argue that the demarcation problem is unsolvable and uninteresting. Karl Popper called this problem the *"central question"* of philosophy of science, and argued that the core property of science is falsiability. In his 1934 book *"Scientific Discovery"* he proposed that scientific ideas can *only* be tested through falsification, newer through a search for supporting evidence. In other words, his goal was not to acquire evidence to prove a theory to be true, but rather to find evidence that would prove a theory to be false. For example, are all swans white? Collecting white swans from all over the world would apparently prove the claim to hold, but finding one black Australian swan would immediately falsify the claim. As long as no non-white swans are found the likelihood of the thesis to be true increases, but it is not proven. If one non-white swan is found, the claim is surely rendered false. Popper believed that scientists should always be skeptical about their theories and seek every opportunity to try to falsify them. One possible way to distinguish between science and non-science was proposed by *logical positivists* who grounded science in observation, whereas non-science was not observational and thus uninteresting. Actually, science always results in theory, not pure objective truth.

3.3 Rationalism

Rationalism was established by Renè Descartes in his 1637 treatise *"Discourse on Method"*. He developed the scientific method that is based on deduction from first principles. His ideas and mathematics influenced Isaac Newton and other scientists on their works. Deduction is the reasoning method in which a conclusion is logically obtained from premises. For example, if the current relative positions of the sun, the Earth and the moon are known, and also how these move with respect to one another, then the date and location of the next solar eclipse can be *deduced*. Of course the philosophical question here is *whether the rational is the real*. The works of Francis Bacon (who promoted the alternative scientific method in which scientists collect many facts from observations and experiments, and then make inductive inferences about patterns in nature), and Descartes have put the foundations of the modern scientific method. Another founder of modern science is David Hume who gave a definite formulation of the *induction method*.

The four main forms of *logical rationality* are:

- Deductive reasoning
- Inductive reasoning
- Abductive reasoning
- Non-monotonic reasoning

Deductive reasoning This reasoning represents the absolutely certain rationality. Antecedents (premises) A_1, A_2 ,..., A_n deductively support (or entail or yield, or imply) a consequent (conclusion) C if and only if C must be true if all of the A_is are true. In other words C is true if and only if the A_is are "true preserving". As indicated in Section 2.4.1 C can be false! It only has to be true relative to the antecedents, i.e., true relative to its context.

Inductive reasoning In this reasoning, also called *probabilistic reasoning*, C is probably true if allA_is are true. This means that unlike deductive inferences, inductive ones do not guarantee the truth of their conclusion. As long as no case is found that the conclusion is not true the likelihood of the thesis to be true increases, but it is not proven. If one case is found that the conclusion is not true the claim is certainly rendered false (as in the example of white swans mentioned in Section 3.2.1 of falsiability).

Abductive reasoning This reasoning, also referred to as *"inference to the best explanation"*, is again probabilistic. From observation P made at time t_1 and a theory T that deductively or inductively entails P, it can be abductively inferred that T must have been the case at an earlier time t_0, i.e., T is an explanation of why P has been observed. The more observations that T explains, the better the theory T is. A different kind of abduction is that where "from observation P_1 at time t_1 and observation P_2 at a later time t_2, it can be abductively inferred that P_1 might have caused or logically implied P_2". Like inductive logic, abductive logic does not guarantee the truth of its conclusions. Actually, abductive inferences are deductively not valid, but they are at the core of the scientific method for developing and verifying theories.

Non-monotonic reasoning This kind of logical reasoning is more *"psychologically real"* than the others. Herbert Simon called this reasoning *"satisficing"* or being satisfied with a reasonable answer to some question rather than an optimal one. In monotonic reasoning (like deductive reasoning), once it is proven that a consequent C follows from an antecedent A, it will always follow. In non-monotonic reasoning, one might infer conclusion C from antecedent A at time t_0, but at a later t_1 he/she might see that this is not true. In this case the beliefs must be revised.

3.4 Empiricism vs. Idealism

Empiricism This philosophical school (originated by the British philosopher John Locke: 1632–1704) advocated that all knowledge comes *a posteriori* from experience. According to Locke, man is born as a blank state *(tabula rasa)* which is subsequently enriched by experiences that in turn become knowledge. This was the base of the so-called the *justification model* of knowledge. Starting with observation one is led to the formulation of a hypothesis which is culminated in a theory about reality and translates into *truth*. It is challenging to specify what is meant by a justification (explanation) model when the thing to be explained cannot be deduced from any law because it has been occurred by chance, or otherwise cannot be perfectly predicted from what is known. Wesley Salmon proposed a model in which a good scientific explanation must be statistically relevant to the outcome to be explained.

Idealism Immanuel Kant in his 1781 book *"The Critique of Pure Reason"* criticized the rational and empirical traditions and attempted a synthesis of both. With regard to the *justification model*, he argued that the step from observation to theory is impossible and the step from theory to truth is highly questionable. Instead, Kant proposed that the correct scientific method began with theory and in the context of the theory in which the observations are made. This means that our presuppositions and theories reflect the way we view the world around us such that no one is objective in his/her understanding of the world and in his/her science. According to Hans Reichenbach (1891–1953), and other philosophers, a distinction must be made between *"context of discovery"* and the *"context of justification"*. Once a hypothesis has been proposed, there are rules of logic that determine whether or not it should be accepted, i.e., there are rules of method that hold in the context of justification. However, no such rules exist to guide someone to formulate the right hypothesis or even hypotheses that are plausible or reasonable.

3.5 Two Other Theories of Confirmation

Besides the falsification theory there are many other theories of scientific confirmation. Here we will outline two of them, namely:

- Logical positivism
- Paradigm model

3.5.1 Logical Positivism

This theory is also called verificationalism, because it accepts only testable statements as meaningful, and rejects metaphysical interpretations. Verificationalism embraces a set of theories of knowledge that combine rationalism, empiricism, and linguistics to ground philosophy on a basis consistent with paradigms from the empirical sciences. It was originally developed by a group of philosophers in Vienna called the *Vienna Circle*, and clearly articulated by A.J. Ayer in his 1946 book *"Language, Truth and Logic"*. Verificationalism is very close to the justification model described above and advocates that for a statement to be true it must be either empirically verifiable, or it must be *tautological* (where the predicate defines) the subject. Ayers expressed his position as follows: "To test whether a sentence expresses a genuine empirical hypothesis, I adopt what may be called a modified verificational principle, for I require of an empirical hypothesis, not indeed that it should be completely verifiable, but that some possible sense experience should be relevant to the determination of its truth or falsehood. If a proposition fails to satisfy this principle, and is not a tautology, then I hold that it is metaphysical, and that being metaphysical it is neither true not false by literally senses". Thus, only the verifiable is considered to be scientific and cognitively meaningful in logical positivism. The unverifiable is unscientific, cognitively meaningless "pseudo statements" (metaphysical, emotive, etc.) not worthy of further consideration by philosophers of science.

Logical positivism has received strong criticism namely:

- Is the verificational principle empirically verifiable, or a tautology? Hence, on its own grounds, the principle is meaningless or senseless.
- It rules out discussion of the nonempirical including God, morality, and even atomic particles.
- While this theory claims to be inductive and objective, it is actually highly deductive and subjective. In other words, it puts the theory and framework of verificationalism upon experience and what is considered to be empirical evidence for supporting a hypothesis is thereby specified *a priori* by the scientist's own empirical experience (hence its subjective nature).

3.5.2 Paradigm Model

This view of science was proposed by the philosopher Thomas Kuhn in his 1962 book *"The structure of Scientific Revolutions"*. He argued that the picture of science developed by logical empiricists such Popper didn't resemble the history of science. He famously distinguished between *normal science*, where scientists solve problems within a particular framework or paradigm, and *revolutionary science*, when a number of scientific irregularities (abnormalities) occur that do not match the paradigm in consideration. He argued that the process of observation and evaluation takes place within a paradigm. A paradigm is a logically consistent "image" of the world, i.e., it does not involve any logical contradictions, and is consistent with observations performed from its point of view. Kuhn rejected the view that it may be possible to isolate the hypothesis being tested from the influence of the theory in which the observations are based, and argued that it is impossible to evaluate competing paradigms independently. The choice between paradigms of a scientist involves setting two or more "portraits" against the world and deciding which is likely more promising. Actually, a paradigm shift will take place when a remarkable number of observation anomalies in the old paradigm have rendered the new paradigm more appropriate and useful. Acceptance or rejection of a paradigm takes place both as logical and social process with the consensus of the scientific community.

In his 1975 book *"Against Methods"* Paul Feyerabend argued that science does not proceed rationally or progressively, but rather irrationally from one paradigm to another. He claimed that no description of scientific method could possibly be sufficiently broad to encompass all the approaches and methods employed by scientists, i.e., no exception-free and useful methodologies exist that govern the progress of science. Feyerabend stated that "the only principle that does not inhibit progress is: *anything goes*", i.e., that there is no such thing as normal science and all scientists should have full freedom to pursue nature as they so desire. There must be no constraints, but rather a complete anarchy as the history of science suggests. This is called "epistemological anarchistic model". For his declaration of epistemological anarchism his detractors named Feyerabend as *"the worst enemy of science"*.

3.6 The Questions "Is Science Philosophy?" and "Could Philosophy Be Scientific?"

Is Science Philosophy?   Empirical or experimental methodology of science is not deductive, but provides highly likely conclusions, and is usually the best we can get. It can be stated that *science is philosophy*, as long as experiments and empirical methods are considered to be *rational* and lead to *truth* [6]. *Scientific realists* argue that science aims at truth and that one ought to view scientific theories as true, approximately true, or likely true. On the contrary, *scientific antirealists* claim that science does not aim (or at least it does not succeed) at truth, particularly truth about un-observables (atoms, electrons, etc.).

Furthermore, instrumentalists argue that scientific theories must be evaluated only on whether they are useful. Their position is that whether theories are true or not is irrelevant, because the aim of science is to make predictions and make possible effective technology. Physics and psychology used to be branches of philosophy, e.g., Newton's book that founded modern physics was entitled: "Principia: Mathematical Principles of Natural Philosophy", while Aristotle and Kant wrote physics books. Albert Einstein as a physicist wrote philosophy, and the philosopher Willard Van Orman viewed philosophy as lying on a continuum with science.

Science cannot be viewed as philosophy if experiments are not considered and do not count as being rational but only as logic. Science is also not philosophy if philosophy is regarded as the search of *universal or necessary truths*, i.e., things that would be true irrespectively of what results are found by science or what basic assumptions are made.

Could Philosophy be Scientific? The question is whether philosophy could or should be more scientific, i.e., more experimental than it is. An area where philosophy can be part of scientific world view is experimental psychology, where philosophers perform scientific/psychological experiments to find out what people who are not philosophers think about certain philosophical topics (although this can be regarded as belonging to cognitive science) [6]. According to Colin McGinn, philosophy is a science, i.e. a systematically organized body of knowledge, and calls it *"ontical science"* that searches for the essences of things using *a priori* methods. As mentioned above, Van Orman considers that philosophy and science belong to the same continuum which means that philosophy is aware of, and makes active philosophical use of scientific results. In this way philosophy can help science in its development and is not restricted to merely examine what others say or how they do philosophy. This kind of philosophy is called *"naturalistic philosophy"*.

3.7 Philosophy of Technology: General Discussion

Philosophy of technology is a relatively young field. During the last two centuries, when it gradually emerged as a discipline, philosophy of technology was mostly interested with the impact of technology on society and culture, rather than with technology itself. It was regarded to belong to the same continuum where social science and humanities belong, and was initially known as *"humanities philosophy of technology"*. On the assumption that technology is the application of science, and that technology is for the good of society, most philosophers believed that there was of little interest. As a result of this point of view, only recently a branch of the philosophy of science has developed that is concerned with technology itself and aims at understanding both the process of designing and creating artifacts, including artificial processes and systems, and the nature of these artifacts. The branch of philosophy of technology involves the interaction of several different fields of knowledge such as philosophy of science, political philosophy, economic philosophy, decision making philosophy, engineering philosophy, and philosophy of design. As a result of this, the field of philosophy of technology is hardly consolidated even in present days. Engineers are focusing on the practice of technology, in the same way as scientists are focusing on the practice of science. Therefore, it could be thought that analytic philosophy of technology amounts to philosophy of engineering. In fact, many of the issues related to design, could be considered that form the subject matter of philosophy of engineering. But, the metaphysical issues (ontology, epistemology, teleology of the artifacts) cannot be included in the philosophy of engineering, which means that philosophy of technology is much broader than philosophy of engineering.

A definition of the philosophy of technology (which aims at inclusiveness rather than delimiting the field) was given by Kaplan (2004) and is the following [10]:

"Philosophy of technology is a critical, reflective examination of the nature of technology as well as the effects and transformation of technologies in human knowledge, activities, societies, and environments. The aim of philosophy of technology is to understand evaluate, and criticize the ways in which technologies reflect as well as change human life individually, socially, and politically".

Historically, the philosophical reflection of technology has its origin in ancient Greece with four dominant aspects namely:

- Technology learns from or imitates nature (Plato, Democritus, Heraclitus).
- Natural entities and artifacts have a fundamental ontological distinction (Aristotle).
- The four causes doctrine, material, formal, efficient and final, of artifacts such as houses and statues (Aristotle).

- The extensive use of technological images in describing the world, e.g., images drawn from carpentry, weaving, ceramics, metallurgy, and agriculture technology (Plato, Aristotle).

In the Roman empire as well as during the Middle ages, the philosophical perspective of technology has not shown a substantial growth. During the Renaissance the philosophical reflection on technology and its impact on society were increased due to a greater appreciation of human beings and their creative activities, including technology. Francis Bacon was the first modern thinker who put forward such reflection with a very positive view of technology that was maintained during the 19th century and the first-half-century of the industrial revolution. During the last quarter of the 19th century and most of the 20th century a criticism of technology was predominated in the philosophical reflection of technology schooled in the humanities or social sciences. The first author who coined the term *philosophy of technology* was Ernst Kapp in his 1877 book *"Eine Philosophie der Technik"*. Most of the authors who wrote critically about technology and its socio-cultural impact during the 20th century were general philosophy thinkers (e.g., Martin Heidegger, Arnold Gehlen, Andrew Feenberg) with a background in social science or humanities. The philosophical view advocated by the above authors was called by Carl Mitcham "humanities philosophy of technology", in his 1994 book *"Thinking Through Technology"*. A form of philosophy of technology that can be thought as an alternative to the humanities philosophy of technology was emerged in the 1960s, and is the so-called "analytic philosophy of technology" which is concerned both with the relations between technology and society, and with technology itself. This field has received increasing attention as documented in the recent works of Scharff and Dusek (2003) [12] and Paul Durbin (2007) [9].

3.8 Fundamental Questions in Philosophy of Technology and Philosophy of Engineering

3.8.1 What Is Technology?

To define technology is not an easy task as it seems to be from a first glance. Therefore, many attempts were made to define technology as more completely and convincingly as possible. A definition can be given through the relation of technology and science. From this point of view Henryk Skolimowski stated in 1966 that technology is something quite different from science. In his own words: "Science concerns itself with *what is,* whereas technology concerns itself with *what is to be"*. A similar definition was also given by Herbert Simon who in his 1969 book *"The Sciences of the Artificial"* stated that "the scientist is concerned with how things are whereas the engineer is concerned with how things ought to be".

The above definitions view technology as a continuous attempt to bring the world closer to the way one wishes it to be, i.e., while science aims to understand the world as it is, technology aims to change the world. But, the question here is "whose wishes concerning what the world should be like are realized in technology?" Unlike scientists, who in many cases are personally motivated in their efforts to understand and explain the world, technologists and engineers are often motivated by the desire to change the world as a service to the society. Therefore, technology is regarded as being *instrumental*, delivering products ordered from "elsewhere", and as means to ends specified outside engineering. Technology is a practice aiming to create "artifacts" and "artifact-based services".

On the basis of the above, there is presently a consensus that technology should be defined using the concept of *"technological system"* which involves materials (hardware), knowledge (rules, guidelines), designers, inventors, software engineers, operators, programmers, customers, distributors, marketers, advertisers, economics scientists, government administrators, and all those involved in a technology. This technological systems approach is more comprehensive than technology which is restricted to either the tools (hardware) or the rules (software), because it includes both together with the people involved in it [8]. Of course the role of engineering (engineers and engineering design) is the dominating factor in the technology development, and this justifies the recent emergence of *"philosophy of engineering"* and *"philosophy of design"* as distinct subfields of philosophy of technology.

3.8.2 The Questions

Similar to the philosophy of science, the questions of philosophy of technology range from metaphysical (ontological) to epistemological, ethical and aesthetics questions. A non-exhaustive list of questions is the following:

- Are there problems of philosophy of mind in technology?
- What are the status and the character of artifacts?
- What is engineering knowledge?
- What is engineering?
- What is a tool?
- What is the purpose of technology and engineering?

- How does uncertainty inherent in technology and engineering differ from scientific uncertainty?
- How is engineering distinguished from artisanship?
- What is creativity?
- Are technological problems self-determined or socially determined?

Regarding the design process some basic philosophical questions are:

- What is design?
- Is there an engineering design method?
- Does design matter, and to what? Why? How? For what purpose?
- What is design methodology, and how it influences the end product of design?

Let us first have a look at the ontology (status and character) of artifacts. Artifacts are man-made objects and those that are of relevance to technology are made primarily for serving a purpose. Of course several by-products, such as waste products, are not generated for a purpose. The same is true in most cases for the works of art. Works of art are the outcome of an intention directed at their creation (and possibly at their intermediate steps), and is questionable whether artists include among their intentions an intention that their work serves some purpose. Of course this issue is a topic of philosophy of art. The creation of an end artifact is achieved via the creation of intermediate products. Whether an artifact is an end product or an intermediate one or a component, or it is produced "for something", and "what it is", is called the *artifact's function*. Many thinkers of technology have suggested that an adequate description of an artifact must include both its status as a tangible physical object and the intentions of the people engaged with it. The concept of functions is very important for the characterization of artifacts, but is also used more widely. For example, the concept of function is also a key concept in biology and in cognitive science.

Historically, engineering has been interpreted as *"devising"*, but the distinction between art, craft, and technology is not clear cut. The Latin root *ars* (art), the German root *kraft* (craft) and the Greek root *techne* (τέχνη) all originally meant the ability or skill to create something tangible (not intangible as e.g. athletic ability). Today, the term art is typically used in visual, performing or literary fields that are called *fine arts* (e.g., the art of writing). The term *craft* is used to the manual ability for manufacturing an object, e.g., the craft of typesetting or dress making. The term *technology* tends to mean the processes and products used in an industry (e.g., the technology of printing, the technology of electronic circuit building, etc.). But, engineering is the activity of performing change via the design and manufacture of artifacts (e.g., the engineering of print technology or the engineering of electronic circuit building). Engineering design differs from artistic design because it is required from the engineer to make quantitative predictions of the behavior and effect (to the individual and the society) of the artifact before its manufacture. Therefore, engineering can be regarded both as a technological and a social field, and should be evaluated not just whether its artifacts work but what is their impact and service to society. Socio-technical systems (power utilities, transport, etc.) involve both artifacts and humans. Moreover, engineering of artifacts has implications to all issues of human life (e.g., sociology, human development, economic wealth, health, psychology, and so on).

Let us now list a number of philosophical questions about design [20]:

- Does design matter? The reply is *yes*.
- To what? To the artifacts which will be different thanks to design?
- Yes, but why? Because it depends on the particular knowledge used. This is an epistemological issue.
- Yes, but for what purpose? For the purpose of changing our life and environment in a desired way. This is a "Teleological" issue.
- Surely yes, but how? How design can be practiced to lead to the desired outcome? By applying the proper methodology. This brings methodology as a subfield of philosophy, where methodology has not necessary to do with a specific method.

As it was already pointed out, artifacts have both a physical nature (geometry, shape, color, structure, size) and a functional nature that allow them to satisfy the desired goals. The purpose of design is exactly to find a proper match between these two, i.e., designers attempt to realize artifacts that have a physical nature which fits well with the intended function characteristics. Good designers take into account the proper function and also try to predict in what other ways the products may be used always respecting the safety requirement.

The question of "what is the nature of technological knowledge" seems not to belong to traditional epistemology because the latter is mainly based on scientific knowledge. This shows that there is one more difference between technology and science. Design is carried out for purposes, which may differ considerably depending on who are involved in designing, constructing, and using the artifact.

Design methodology is the reflection on design practice in terms of the *"how to do it"* issue. The studies of design methodologies have revealed that well defined methods play a limited role in design practice. In case well defined methods are used, care should be taken to assure that their built-in assumptions hold. If these assumptions are not met the employment of the method may well be problematic despite the fact that at first glance seems to be appropriate for getting the desired outcome.

It should be pointed out here that "philosophy of design" is not the same as *design philosophy*. It is disciplinary similar to philosophy of science or philosophy of technology, whereas *design philosophy* has to do more with the philosophical study of the design method. Over the years a vast amount of definitions of *design* and *design process* have been presented which cannot be unified in general concept, because different authors were using the same terms or concepts quite differently.

Chronologically, the following overlapping topics of research into design have been identified [15]:

- The management of the design process.
- The structure of the design problems.
- The nature of design activities.
- Reflection on the fundamental concepts of design.
- Knowledge concerning the environment where designing is performed.
- The knowledge required for designing (i.e., knowledge about objects and design processes).

Taken together these topics provide the different ways in which design researchers have studied design activities, and proposed theories and models for representing them. To clarify a design theory the so-called *"critical analysis of design methods"* has been developed. This is a field that aims to investigate the relationships between various concepts and theories associated with design, as well as their underlying assumptions. However, this is not an easy task, because even a single author has defined a concept to have a particular meaning and then used it later in the same publication with an intelligently different meaning.

Other important questions that have to be addressed by philosophy of design are the following:

- What distinguishes design (engineering, architectural, hardware, software, etc.) from other intellectual endeavors such as science or technology?
- How are the concepts of design and artifact related?
- How can ontology and epistemology help to explain the apparent fact that designers can know or anticipate the properties of the artifact to be designed?

- Can results from the philosophy of design be put to use in practice such that to get better artifacts, better design methods, better ways of employing computers, etc.?

The common feature in all these issues is that they are aspects of design, and that insights about them can be obtained by rational and critical reflection rather than empirical observations.

A simple way to proceed to a *"critical analysis"* of design methods is to take a *meta-theoretical* perspective, based on several levels of abstraction that provide a hierarchical classification, which is relatively independent of the domain based meaning or interpretations of each theoretical element. A convenient hierarchy of meta-theoretical levels of abstractions referred to design is the following (in bottom-up order) [15]:

- *Direct perception of realities* (spatial, visual, hearing, linguistic perception, etc.).
- *Description of objects* (including processes, systems, software tools and programs, etc.).
- *Behavior of elements* (description of the behavior of objects, processes, systems, etc.).
- *Mechanisms of choice* (the way that choices are made between objects, systems, processes etc., and how the choices are evaluated).
- *Design methods* (the level at which design methods and techniques are described).
- *Design process structure* (this level includes the theories about the design process structure, and the influences of domain, culture, artifact type, etc.).
- *Internal processes of designers* (this level involves the description of theories about the cognition and reasoning of individual designers or teams of designers, and of cultural design effects on designers).
- *General design theories* (this level describes the design theories used for describing the whole activity of designing the desired objects).
- *Epistemology of design* (this is the level where analyses and discussions about the critical study of the nature, limits, criteria or validity of design knowledge are performed).
- *Ontology of design* (this is the level where the ontological issues of design theory and the activity of design are studied).

Abstractions from the *"description of objects level"* to the *"ontology of design level"* provide theories and patterns of concepts at lower levels. Abstraction at higher level yields assumptions for the preceding levels. The above hierarchical taxonomy of abstractions coincides with the hierarchical levels of creativity which have been studied in psychology. Therefore, replacing the word *"design"* by the word *"creativity"* the above design abstraction levels, can also provide the creativity-of- design levels. Both interpretations of this hierarchical taxonomy can contribute towards the reduction of confusion, semantic polymorphy, and lack of coherency in design theory.

3.9 Concluding Remarks

In this chapter we provided an account of philosophy of science and philosophy of technology as a preparation for the discussion of philosophy of robots. The questions "what is science", "what is the distinction of science from non-science", and the philosophical principles of rationalism, logic reasoning, empiricism, and idealism were addressed. Other aspects studied in the chapter are: (i) the logical positivism and the paradigm model of confirmation, (ii) the questions "whether science is philosophy" or "philosophy can be scientific", (iii) the questions "what is technology", "what is engineering", and "what is design and philosophy of design", and (iv) a hierarchical meta-theoretical taxonomy that facilitates the critical analysis of design. The difference between philosophy of science and philosophy of technology is that the former deals with the ontological question "what is", whereas the later is concerned with the questions "what is to be or should be", i.e., with the issue of changing the world at will.

In closing this section, we list a few big names in the field of philosophy of science, indicating their principal achievements.

Aristotle (384–322 BC): He is considered as the father of both science and philosophy of science. In addition to mathematics, logic and epistemology, he studied extensively the topics known today as physics, astronomy, chemistry, and psychology.

Francis Bacon (1596–1650): He developed the scientific method which is based on gathering many facts from observations and experiments, and then inductively infer about patterns in nature.

Rene Descartes (1596–1650): He developed and promoted the "deductive" reasoning method from first principles (the so-called rationalism).

Pierre Duhem (1861–1916): He defended an extreme form of empiricism. He promoted John Locke's "empiricism theory", arguing that we cannot draw conclusions about the existence of unobservable entities conjectured by theories such as atoms and molecules.

Karl Popper (1924–1194): He promoted the falsiability method, stating that falsiability is the proper scientific approach, i.e., scientists must always question their theories, regarding them with skepticism and seeking every opportunity to try to falsify them.

Thomas Kuhn (1922–1996): He argued that the picture of science developed by empiricism did not resemble the history of science. According to him, science is distinguished in *normal science* (where scientists solve puzzles within a specific paradigm or framework), and *revolutionary science* (when the paradigm gets overturned).

References

[1] P. Godfrey Smith, Theory and Reality: An Introduction to the Philosophy of Science, Chicago/London: The University of Chicago Press, 2003.

[2] A. Rosenberg, Philosophy of Science: A Contemporary Introduction, New York/London: Routlege: Taylor and Francis Group, 2011.

[3] J.A. Cover, M. Curd and C. Pincock, Philosophy of Science: The Central Issues, New York, NY, U.S.A.: W.W. Norton & Co., 2012.

[4] S. Okasha, Philosophy of Science: A Very Short Introduction, Oxford/New York, Oxford University Press, 2002.

[5] Wikipedia Book: Philosophy of Science www.wikipedia.org/wiki/Book: philosophy-of-science [Free Access].

[6] W.J. Rapaport, Philosophy of Computer Science: An Introductory Course, Science Teaching Philosophy, 28(4), 2005, 319–341.

[7] Philosophy of science, Encyclopedia Britannica (www.britannica.com).

[8] V. Dusek, Philosophy at Technology: An Introduction, Malden, MA, U.S.A.: Blackell, 2006.

[9] P. Durbin, Philosophy of Technology: In search of discourse synthesis, Techné, 10(2), 2007.

[10] D.M. Kaplan (ed.), Readings in the Philosophy of Technology, New York, N.Y., U.S.A., Free Press, 1972.

[11] C. Mitcham and R. Mackey (eds.), Philosophy and Technology, Lanham, MD, U.S.A.: Rowman and Littlefield, 1972.

[12] R.C. Scharff and V. Dusek (eds.), Philosophy of Technology: The Technological Condition, Malden, MA: Blackell, 2003.

[13] M.J. de Vries and A. Tamir (eds.), Shaping Concepts of Technology: From Philosophical Perspectives to Mental Images, Dordrecht: Kluwer, 1997.

[14] N.P. Suh, The Principles of Design, Oxford/New York: Oxford University Press, 1990.

[15] T. Love, Philosophy of design: A meta-theoretical structure for design theory, Love Design & Research, 2000 (www.love.com.au).

[16] C. Mitcham, Thinking Through Technology: The Path Between Engineering and Philosophy, Chicago, U.S.A.: Chicago University Press, 1999.

[17] A. Meijers (ed.), Philosophy of Technology and Engineering Sciences: Handbook of the Philosophy of Science, New York: Elsevier, 2009.

[18] Philosophy of Technology, Stanford Encyclopedia of Technology (http://plato.stanford.edu/entries/technology).

[19] H. Skolimowski, The structure of thinking in technology, Technology and Culture, 7, 1966, 371–383.

[20] M.J. de Vries, Design matters, and so does philosophy of design, The J. of Design and Technology Education, 8(3), 2003, 150–153.

4

Robot Cognition

Thoughts without content are empty,
intuitions without concepts are blind.
Immanuel Kant

Behavior, cognitive, and other personal factors,
and environmental influences all operate
interactively as determinants of each other.
Albert Bandura

Cognition reigns but does not rule.
Paul Valery

4.1 Introduction

Historically, *cognitive science* has emerged with the so-called *behaviorism approach* (stimulus-response concept), and then proceeded to the so-called *cognitive revolution* (content-bearing internal states concept) that studies the structure of the internal information processing. Perhaps the first philosophical issue of concern in cognitive revolution was the issue of architecture. Given that inner representational states offered a new point of view to cognition, the question of what processes operate on them and what they are, has been raised.

With regard to cognitive robotics, the main question is *"why do we need cognitive robots?"* The answer is two-fold. First, new technologies enable the design of affordable, safe, and efficient robots. Second, the modern societies bring the need to have robots that work together with humans in real environments as assistants, teammates, care-takers, and companions. Two other important questions in cognitive robotics are: *"can a robot learn as a child?"* and *"does the body have an impact on how we think?"* The first question is studied in the *"developmental robotics"* subfield, and the second in the framework of *"embodied cognitive robotics"*.

The purpose of this chapter is to discuss several fundamental philosophical aspects and questions of both human cognition and cognitive robotics. Specifically, the chapter:

- Addresses the ontological questions "what is cognition" and "what is cognitive robotics".
- Provides the basic historical landmarks of cognitive science and cognitive robotics.
- Presents and discusses a number of issues and questions of "philosophy *in* cognitive science", "philosophy *of* cognitive science", and "philosophy of cognitive robotics" (cognition, perception, embodied cognition, developmental robotics, socialized robotics).
- Gives some technological elements of cognitive robots based on the cognivist approach, the emergent systems approach, and the hybrid cognivist-emergent systems approach.
- Outlines a set of basic requirements for the design of cognitive robots.

4.2 What Is Cognition?

Cognition is, in general, defined as the process of knowing, including the aspects of perception, awareness, reasoning, judgment, and intuition [1, 2]. The word cognition comes from the ancient Greek word γνώσκω (gnosko = know or learn) and the derivative Latin word *cognosco* (co = with, gnosko = know) which is interpreted in the broad sense to mean conceptualize or recognize. *Cognitive science* is the science of mind, and nowadays speaking about cognition we are referring to the mind as an information processor, i.e., a system that acquires, uses, and transforms information [3]. Cognitive science interfaces with philosophy in three ways:

- It borrows functions of philosophy of science (such as explanation, reduction, logical reasoning) and uses them in cognitive processes.
- It applies metaphysical philosophy to study questions about the mind (such as the nature of perception, consciousness, mental representation, emotions, conscience, etc.).
- It attempts to construct the foundations of a theoretical framework and tools for getting a science of the physical basis of the mind off the ground, the so-called "theoretical cognitive science" that has one foot in cognitive science and the other in philosophy.

Cognition, however, is not just restricted to the kind of explicit knowledge and rational thinking or philosophical reasoning, but also includes subconscious, intuitive, and effective feelings and experiences which are also based on information processing. This means that cognition includes the entire functioning of the brain at the higher level, without the details of the neurophysiological brain anatomy, although it may get inspiration from these details if they provide insight about the higher-order functions. The focus of cognition is in the function of the brain and its components: "What, how, and why it does what is doing?"

In psychology, the study of behavior is distinguished in:

- *Cognitive behavior* (how we know the world).
- *Affective behavior* (how we understand the world via feelings and emotions).

Currently, much effort in psychology is focused on *cognitive psychology* of emotion including one's awareness, and one's own strategies and methods of cognition known as *"meta-cognition"* which includes *"meta-memory"*. However, there are many psychologists who do not accept that cognitive processes are functions of the brain. A theory of cognition is not necessary to make reference to the brain, because it may partly describe behavior via information flow, information function, and information processing.

Recent studies in theoretical cognitive science attempt to bridge the gap by using cognitive examples to illuminate how the brain implements these information processing functions, and simulating them via artificial intelligence techniques. The cognitive development of infants was first studied by *Jean Piaget* [4] who formulated a theory that describes the stages children pass through during development. This theory placed the foundation of the now called *developmental psychology*. The principal postulates of this theory are that children construct their own knowledge in response to their experiences, and that they are capable to learn alone, without the help from other children or adults. This theory was inspired by the *empirical movement* of *John Locke* [5] who presumed that humans are born without knowledge like a *blank slate* (tabula rasa) and are formed by experiences in the society and environment due to their capacity to do *"abstract symbolic reasoning"*. Developmental psychology was originally concerned with infants and children but in nowadays encompasses the entire human life span which includes other life periods of great change such as adolescence and aging.

4.3 What Is Cognitive Robotics?

Cognitive robotics is the intersection of robotics and cognition (or cognitive science). Robotics is the science, technology, and engineering of robots. Today, the word *robot* is typically referred to any machine that works automatically or by remote control to perform tasks normally performed by humans. The word *robot* was firstly used by Karel Capek in his 1920 play called "Rossum's Universal Robots" (R.U.R). It comes from the Czech world *"robota"* that means forced or slaved labor. The term *robotics* was coined in the 1940s by the science fiction writer Isaac Asimov who formulated the well known "Asimov's Laws of Robotics":

- *First Law* A robot may not injure a human being, or, through inaction, allow a human being to come to harm.
- *Second Law* A robot must obey the orders given it by human beings except where such orders would conflict with the First Law.
- *Third Law* A robot must protect its own existence as long as such protection does not conflict with the First or Second Law.

Asimov's book *"I Robot"* explores the interplay between these laws which constitute the foundation of ethics in robotics. As said in Section 1.1, the first industrial robot that was put in operation (1961) is the robot *Unimate* designed by George Devol and Joseph Engelberger. This robot plucked hot automobile parts out of a die-casting machine and quenched them in water. Industrial robotics saw a revolution in the 1960s and 1970s. The robots replaced humans for many repetitive jobs performing through preprogramming *dull, dangerous* or *dirty* work (the well known *3D work*). Even today there are still many robots of this type in factories, but the trend is to have more intelligent and autonomous robots that have cognitive capabilities such as perception, reasoning, learning and interaction with humans in real non-structured environments. These robots can be characterized as *3C robots* (Clever, Creative, and Charismatic), i.e. cognitive robots. One way to get cognitive robots is to use, embed and implement into them the most accurate models and the best results of cognitive science. This means that *cognitive robotics* can be seen as application of cognitive science to robotics, or, equivalently, as doing robotics informed by cognitive science. Clearly, cognitive robotics can in turn be used for testing in practice cognitive science models and theories. For example, we can implement two available theories or models that explain the way humans perform some cognitive processes in a robot, and see which theory enables the robot to imitate better the human performance. An important class of cognitive robotics is the so-called *developmental robotics* which is mainly based on the principles and results of developmental psychology. Unlike evolutionary robotics that operates on phylogenetic time scales and populations of many individuals, developmental robotics operates on short (ontogenetic) time scales and single individuals or small groups of individuals. Developmental robotics is also known as *ontogenetic robotics* or *epigenetic robotics*. In developmental robotics, learning is cumulative and has a progressively increasing complexity. It is autonomously acquired from self-exploration of the world and social interaction.

4.4 Historical Remarks of Cognitive Science and Cognitive Robotics

4.4.1 Cognitive Science

Cognitive science has its historical origins in the intellectual activity of ancient Greek philosophers (Socrates, Plato, Aristotle) and the philosophers of the Renaissance period, but its emergence as a collaborative endeavor of philosophy, psychology, artificial intelligence, linguistics, neurosciences, anthropology and related disciplines has occurred in the 1950s. One of the recognized founding contributors of cognitive science in this multidisciplinary view is *George Miller* [20]. By the 1970s disciplines like philosophy, sociology, and anthropology were making important visible contributions, and recently neuroscience and cognitive neuroscience once again became central contributors.

Socrates approach was that genuine knowledge can only be obtained if there are valid concepts, i.e., concepts for which we can give a correct definition, and that the only way to get a correct definition is through dialogue and continuous revision of the definition.

Plato extended further this view and argued that our sensory experiences consist only of floating images which at best provide a misleading picture of the nature's reality. The only realities that are unchanging are abstract concepts (forms), mathematical entities (numbers and perfect geometric figures), and immaterial souls. Thus, reality is not residing in the specific objects we perceive but in the abstract forms that the objects represent.

Aristotle broadened this perception theory by proposing the "theory of mind" (Greek νούς = nous) often rendered as "Intellect" (or "reason") that is part of the soul (Aristotle, DeAnima). He distinguished practical intellect from theoretical intellect, and argued that perception is a case of interaction between objects capable of acting (that have a body), and capacities capable of being affected (Aristotle, Nicomachean Ethics). Perceived objects are configured in the mind by forms (concepts). Aristotle also argued that the route to knowledge is via empirical evidence obtained from experience and observation, and by this put the foundations of modern psychology.

Galileo distinguished sharply primary incidents (e.g., size, motion) which are logically inseparable from body substance, and secondary incidents which are only in the mind of the observer.

Descartes asserted, in addition, that all the things in the world are accessed indirectly through our ideas. The world includes all the contents of the *mind* (images, concepts, perceptions, beliefs, intentions, etc.) which resides outside the physical human body.

Bertrand Russels provided a theory of knowledge which analyzes the differences that may exist between various cognitive relations (attention, memory, sensation, imagination, etc.) and explained how cognitive data (perceptions, concepts, etc.) can become elements of knowledge. He distinguished between direct and indirect knowledge, i.e., between knowledge by acquaintance and knowledge by description.

Immanuel Kant introduced the concept of *a priori* knowledge arguing that our understanding of the external world is based not merely on experience but on both *perceived* knowledge and an *a priori* knowledge. In this way offered a non-empirical criticism to the rationalist philosophy of the renaissance. He stated that both immediate representations (intuition) and mediated representations (general features of things) are essential for getting complete knowledge of the concept.

Behaviorism   This theory started as an adventure for experimental psychology in the early 1900s, and is based on the belief that a science should admit only entities, states, operations etc. that might be reduced to or represented by publicly observable entities, states or operations. Behaviorism was associated with several names: empiricism, operationalism, verificationism and logical positivism, and excluded private (not publicly observable) entities or states such as thoughts, feelings, consciousness, dreams, and so on. Classical behaviorists include *J. Watson* [6] and *B.F. Skinner* [7]. The basic paradigm underlying behaviorist theories and observations is the *stimulus-response* (**S-R**) paradigm (experiment) which involves the following stages:

- *Stimulus* Repeatedly expose a person (or animal) to an event that can be perceived (e.g., a light, a sound, or readable word).
- *Response* Precisely observe each time the behavioral reaction (a movement made, something said, a sound made, etc.).

- *Relation* Try to find the relation or correlation between stimuli and responses (i.e., how far the same stimulus tends to produce the same response under the same conditions).

For a behaviorist the mind is not much more than an enormous collection of S-R pairings. According to *Quine* [8] the "meaning" of linguistic expressions is not mental or non-physical entity, but, to a good approximation, simply sets of stimulus conditions. Two classical examples of behaviorism are the *learning by conditioning* (e.g., Pavlov's dog feeding shortly after the sound of a bell), and the *learning by operand conditioning* (reward or punishment based learning).

Behaviorism is of great historical interest because it was the context for the cognitive revolution which emerged from the realization that any adequate theory of human and animal mentality requires the identification of representational states between stimulus and behavioral response. These states would be theoretical, not simply reducible to constructs of observables.

Cognitive revolution Behaviorism was an exciting field but by the mid 1990s psychologists recognized its weaknesses. Cognitive revolution reopened interesting co-operations between U.S.A. and European researchers in psychology. It brought about a renewed legitimacy of discussion and theorizing about certain kinds of mental or cognitive states, particularly *content-bearing* states (e.g., beliefs, desires, etc.) or more general states over which operations such as inference could be carried out to make plans or decisions. Chomsky pointed out that defining psychology as the science of behavior is like defining physics as the science of meter reading. For psychology to succeed mental concepts would have to integrate and explain the behavioral data. The cognitive revolution has influenced the study of the mind in three directions:

- *Psychological* (Tolman, O'Keefe, Dostrowski)
- *Linguistic* (Chomsky)
- *Mathematical* (Turing, McCulloch, Pits, Newell, Simon, Hebbs)

Edward Tolman **(1948)** demonstrated complex navigation behavior in rats that could be explained not by stimulus-response mechanisms, but by cognitive maps (complex representative states) built by rats [9, 10]. O'Keefe and Dostrowisky appeared to find Tolman's cognitive maps in the brain.

Noam Chomsky **(1957)** argued convincingly that no purely stimulus-driven mechanisms could possibly learn the structure of language, and that language learning seems to require at least some innate cognitive representational structures which circumscribed possible grammars that were then selected from by exposure to linguistic structure [11, 12].

Alain Turing (**1936**) developed the theory of computation and coined the Turing test for asserting whether a system could be regarded as intelligent or not. He stated that hardware is the brain and software is the mind [13, 14].

McCulloh and Pitts (**1943**) developed the artificial neuron concept and attempted to show how neuron-like units acting as logic gates could be interconnected such that to be able to perform computations and learning processes [15].

Allen Newell and Herbert Simon (**1976**) proposed the "physical symbol system" concept by considering the physical patterns as symbols, combining them into structures (expressions), and manipulating them (using processes) to get new expressions. They stated that "A physical symbol system has the necessary and sufficient means for general intelligent action", thus providing a position in the philosophy of artificial intelligence [16, 17].

Donald Hebb (**1949**) connected the biological function of the brain as an organ together with the higher-level operation of the mind. He explained the adaptation of neurons in the brain during the learning process and developed the theory of associative (correlational) learning known as *Hebbian learning*. In Hebbian learning simultaneous activation of neuron cells leads to visible increases in the synaptic strength between those cells. His results opened the door for the creation of computing machines that are able to mimic the biological processes of living nervous systems [18].

George Miller (**1956**) studied the structure of internal information processing system and proposed that memory limitations (short memory) can be faced by recoding information into chunks, mental representations that require mental procedures for encoding and decoding the information [19]. In a 1978 review report compiled for the *Sloan Foundation* he argued that at least six fields are contributing to cognitive science, namely: philosophy, psychology, linguistics, neuroscience, computer science, and anthropology. He constructed an hexagon in which each line connecting two disciplines represents an area of interdisciplinary study that was well defined in 1978 and involves the tools and methods of the two disciplines linked together (Figure 4.1) [20].

The link of Psychology and linguistics is the "psycholinguistics", the link of anthropology and neuroscience represents the studies on the evolution of the brain, the link of computer science and neuroscience (modeling of brain functions) stands for *"cybernetics"*, and so on. Later a seventh discipline was added namely *"education"* leading to 31 fields of cognitive science (each node of the respective hexagon is linked to all other nodes). Due to that the cognitive science in the above senses comprises from several subfields that have their

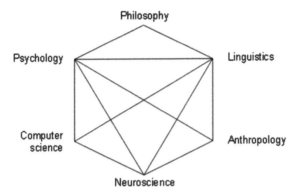

Figure 4.1 Cognitive science in 1978 according to Miller.

Source: http://www.cs.princeton.edu/~rit/geo/Miller.pdf

own independent development, many authors (now including Miller) speak about *cognitive sciences* (plural) instead of simply cognitive science.

Summarizing, we can say that the cognitive revolution has led to the general adoption of the view that complex behaviors are controlled (at least in the larger part) by inner representational states that carry *contents,* and driven by *processes* such that the system can make plans and solve problems. Both of these topics, i.e., the contents carried by representational states, and the nature of driving cognitive processes *("cognitive structure")* have been in the core of the philosophy of cognitive science.

4.4.2 Cognitive Robotics

The history of cognitive robotics has naturally its origin in the early sixties with the first efforts to develop industrial, mobile robots, and humanoid robots pocessing human-like cognitive capabilities of one or the other form or level [21].

In **1963** the *Rancho Arm* was put in operation. This is the first computer controlled robotic arm, a prosthetic arm designed to aid the handicapped, and was used at Rancho Los Amigos Hospital (Davney, California).

In **1969** the first truly flexible arm, known as the *Stanford Arm* was developed at Stanford AI Lab by Victor Scheinman, which soon became a standard and is still influencing the design of today robotic manipulators.

In **1970** which is considered as the starting year of cognitive mobile robots, the mobile robot *Shakey* was developed at the Stanford Research Institute controlled by intelligent algorithms that observe via sensors and react to their own actions. Shakey is referred to as the first electronic person, with its name coming from its jerky motion.

In **1979** the Stanford Cart (originally designed in 1970) was rebuilt and equipped with more robust 3D vision that allowed more intelligence and more autonomy.

The decade **1980–1989** was dominated by the development of advanced Japanese robots, especially humanoid robots (WABOT-1, WABOT-2, Honda's ASIMO), which were capable to walk, or play a musical instrument, or cooperate with humans. These robots, especially ASIMO, are still improved and enhanced with more cognitive and autonomy capabilities.

The decade **1990–1999** is marked by the development of truly behavior-based cognitive robots. Several generic intelligent robot control architectures were developed such as the *"augmented finite state machine* (**AFSM**)-*based subsumption architecture"* of Rodney Brooks or the *"motor schemas architecture"* of Ronald Arkin (1989) [24, 25]. Many other cognitive control architectures were developed in the 2000's up to the present time, especially for the design of socialized (anthropomorphic and zoomorphic) robots, e.g., the *"deliberative-reactive architecture"* of Brian Duffy (1999) [26, 27], and the "**SSS** *(servo-subsumption-symbolic architecture)"* of J. Connell (1992) [28]. Brooks cognitive robot (named **COG**) was an upper-torso humanoid which was used as a platform to generate intelligent behaviors and has widely inspired a number of government and commercial humanoid development projects up to now.

An important class of cognitive robots is the class of *socialized* robots which gave rise to deep research on the *embodiment* or *embodied cognition* issues [26]. Embodiment in robotics refers to the question whether it is *"IN-World"* or *"ON-World"*, a terminology used by Brooks for implementing his subsumption architecture to autonomous mobile robots. This is in agreement with the philosophical view of Heidegger that "we function in the world simply by being part of it" [29]. The embodied cognition philosophers ascribe is the belief that the embodiment of an organism simultaneously constrains and prescribes the types of cognitive processes that are available. The basic distinction between IN-and-ON world embodiment is the concept of robot *adapting* at a *macro* or *micro* level to its environment or not. Smithers [30] argues that we function "ON-World" (not "IN-World") stating that "the agent is directly involved in bringing about the world it experiences, rather than being an external observer of it able to act on it". It should be remarked that *embodied cognition* is different from *embedded cognition*, although they both belong to *situated cognition* [31]. A general view of the three main ingredients (sensing, reflex, motion) of cognitive robots is shown in Figure 4.2, and a general cognitive robot hierarchical architecture in Figure 4.3.

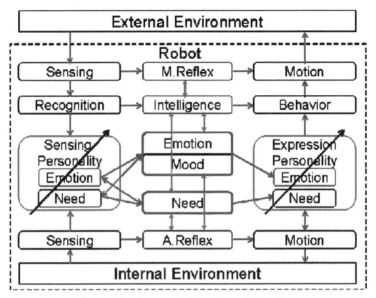

Figure 4.2 A general view of cognitive robots.

Source: http://www.takanishi.mech.waseda.ac.jp/top/research/we/we-4r/img/info_e.gif

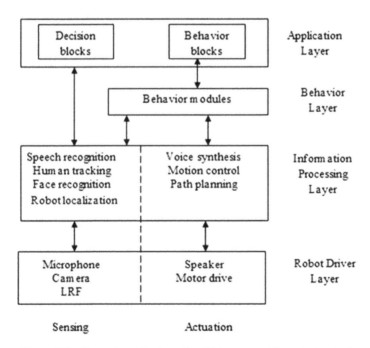

Figure 4.3 General architecture of multi-layer cognitive robot control.

4.5 Philosophical Questions about Cognitive Science and Cognitive Robotics

4.5.1 Cognitive Science

The big question here is *"what philosophy could and should contribute to cognitive science?"* How can this contribution be best made, through understanding cognitive processes by what they do (i.e., applying functionalism), or in terms of their structure or the underlying mechanisms? This is still a debate in the philosophical approach to cognitive science with many different arguments and answers. Andrew Brook [32] has presented the view that to make progress in these philosophical issues one needs to follow two distinct avenues:

- Philosophy *in* cognitive science
- Philosophy *of* cognitive science

The former includes the studies performed for the mind and the language that have also been studied by other approaches such as behavioral experiments and theoretical linguistics. The latter is a meta-study concerned with what others do, rather than doing cognitive science, i.e., it studies cognitive science. Both avenues offer important approaches.

Philosophy in cognitive science   It does not perform experiments, it is not computational, and it seldom does any of the things that other cognitive scientists typically do. Actually, it does *thought experiments* which provide imaginative results rather than formal justification. Of course, thought experiments were also used in science, e.g., physics (Galileo, Schrödinger, Einstein) and psychology, but it is true that they play a stronger role in philosophy than in any other area of cognitive science.

The thought experiments typically involve the following functions [32]:

- *Speculative hypothesis generation* This is done by the application of the imagination to some groups of facts.
- *Integrative interpretation* The activity of interpreting tested hypotheses which provide the result and theories offered by philosophers.
- *Exploration of thought experiments* This involves imagined manipulations of imagined scenarios. Philosophers claim that using them one can reach substantive conclusions.
- *Analysis of concepts* Philosophical work in cognitive science was mostly devoted to clarifying concepts and making recommendations about how concepts should be used.

Thought experiments are quite different from hands-on experiments, but they can do the following:

- They can identify specific examples of a phenomenon being examined.
- They tell us what we initially take something to be like, and help us in deciding what we take something to be like.
- They are often central to hypothesis generation.
- The can often play a significant role in hypothesis testing and elimination.

Actually, philosophy *in* cognitive science is *part* of cognitive science.

Philosophy of cognitive science This view of philosophical consideration of cognitive science raises important issues about guidelines and settings for doing good cognitive research, as well as important meta-questions about the goals of a science like cognitive science in the first place.

Some representative questions that can be asked about cognitive science are:

- Can the various approaches to cognition be integrated into a single unified approach? If yes, how?
- Can studies carried out in a laboratory reveal how cognition works in more real settings, or cognition must be studied *"in the wild"* to achieve such understanding?
- Is human cognition computational in anything like the way any computer (serial or parallel) works?
- Do cognitive models (e.g., SOAR, EPIC, ACT-R, ICARUS, etc.) merely fit the behavioral data or can they capture the actual structure of human cognition?
- What type of explanations is best for cognition, particularly in the light of the massive complexity and interconnections featuring the human brain?
- What do we need to confirm or refute a theory or model of cognition?
- Can general philosophy be applied on cognitive science to determine what kind of science is and what works on it?
- What methods do or should philosophers use to carry out such work?
- What is the role of normative ethical investigations in cognitive science?

4.5.2 Cognitive Robotics

The first big question in cognitive robotics is *"why do we need cognitive robots?"* This question involves several particular questions referred to cognition, perception, embodied cognition, developmental robotics, and social robots. Some of these questions (non-exhaustive) are the following.

Cognition

- What makes cognition possible?
- Can robots develop cognition through sensory-motor coordination?
- Can robots get the ability to communicate with humans using natural language?
- Can robots get the ability to form a mental representation or a model of the user?
- Can robots get the ability to execute efficient, coupled real-world action-taking and the capacity to improve such actions using decoupled off-line reasoning?
- How much human programming is needed to support the learning process?
- Can robots get the ability to have a sense of awareness?
- Can the progress in robot's behavior and success be objectively quantified?
- Can cognitive robots be dangerous?

Perception

- Can robots get the ability to perceive the world in a similar (or better) way to humans?
- How human's sensory systems are working?
- Why are susceptible to illusions?
- How do perceptual states acquire meaning and content?
- How perception is guided?
- What methods do cognitive scientists and cognitive robotocisits use to study perception?
- Is it possible to build an artifact (machine, robot, etc.) that perceives as we do?
- Which structures in the brain perform which perceptual functions?
- Can an artificial *being* (that is able to perceive) have a mind? If yes can we tell whether one does?
- How do one's perceptual experiences compare to those of other people?

Embodied cognition

- Why cognitive tasks such as remembering, are facilitated and performed more effectively by using our bodies or part of our environment to off-load storage, and simplify the nature of the cognitive processing?
- Do body gestures facilitate not just communication but language processing itself?

- Why particular ways in which an organism is embodied limit and prescribe the types of cognitive processes that are available to it?
- Why without the involvement of the body in both sensing and acting, thoughts would be empty and mental affairs would not exhibit the characteristics and properties they do?
- Is embodied cognition constructive? That is, are cognition concepts and categories actively constructed and not merely apprehended wholesale from an observer-independent environment?

Developmental robotics

- Could there be laws governing developmental robots?
- Could there be a theory?
- Could developmental psychology results and human developmental mechanisms or process of thought be used in robotics to increase the learning ability and the adaptation ability of developing robotic systems? How?
- Why do we need developmental robots?

Social robotics

- What can social robots (anthropomorphic, zoomorphic) do?
- Why build androids?
- What is the true potential of social robots?
- Can social robots get the ability to use attention and emotion to exhibit human-like interactivity?
- Can developmental learning be used in creating social robots?
- What kind of reward or punishment is best for a child-like developing social robot?

The above questions about cognitive robots are more philosophical than empirical, and much research is ongoing to find convincing answers or improve existing ones. Many philosophers and robotics thinkers have noted that actually no single research method can answer all these questions about the nature of cognition and its application to create cognitive robots with capabilities truly similar to those of humans (e.g., thinking, learning, feeling, awareness, moral behavior, body-brain-environment interaction, etc.). Representative works, where several questions of the above type are addressed at a technical level, with descriptions of corresponding robots, are described in [33–42].

4.6 A Technical View of Cognitive Robots

In terms of their features and capabilities cognitive robots may be defined as robots that exhibit effective behavior through perception, embodied cognition, deliberation, communication, anticipation, action, adaptation, and self-reflection. Of course, currently not a single robot possesses all the above capabilities, but most of them are designed to learn, reason from experience, to improve their performance, to respond intellectually to situations they never encountered before, to identify potential problems in performing a task, and to know when new information is required to complete the task. Among the many different approaches for studying and building cognitive robots three approaches that are quite popular and offer good generic paradigms are the following [43]:

- *Cognivist approach* which is based on symbolic information representation and processing.
- *Emergent systems approach* which embraces connectionist structures, dynamic structures, and enactive systems.
- *Hybrid cognivist-emergent systems approach* that combine the best features of the previous two approaches.

4.6.1 Cognivist Approaches

These approaches are based on the philosophical view that *"cognition is a kind of computation"* defined on symbolic representations. These representations are instantiated physically as computer codes, and the resulting behaviors of cognitive systems are causal consequences of formal operations performed on these codes [44]. Other names used for the cognivist approach are "symbol manipulation cognition", and "information processing cognition" approach. Typically, the design goal of cognivists is to create systems that reason symbolically on the basis of these representations such that to acquire the desired goal-directed adaptive, anticipatory, and self-reflective behavior. A key feature of cognition system creation using symbolic representation and computation is that semantic knowledge can be directly accessed, understood and interpreted by humans, and that it can be embedded directly into and extracted directly from the system. Of course it should be remarked that this approach is successful when the conditions under which the system operates are not far away from those assumed in formulating the representations, otherwise there may be a large "semantic gap". Cognivism follows the philosophical view of positivism (verificationalism) developed by the *"Vienna Circle"*, according to which "the world we perceive is isomorphic with our perceptions of it as a geometric environment" [45]. Cognivism is closely related to Newell's and Simon's *"Physical Symbol System"* approach to artificial intelligence. A physical symbol system consists of a set of entities, called symbols, which are physical patterns that can occur as components of another type of entity called an expression (or symbol structure). A physical symbol system is a machine that produces over time an evolving collection of symbol structures. Such a system exists in a world of objects wider than just these symbolic expressions themselves. The physical symbol system is grounded on two assumptions [16, 46]:

- A physical symbol system has the necessary and sufficient means for general intelligence action.
- The solutions to problems are represented as symbol structures, and the intelligence in problem solving is exercised through search and progressive modification of the symbol structures until a solution structure is obtained.

Four representative examples of cognivist systems are the following:

- A model-based system that uses description logics (based on predicate calculus) and reasons about high-level concepts (e.g., spatio-temporal events) [47].
- An adaptable system that observes and interprets human activity for dynamically configuring its processing to deal with the context of its operation [48].
- A cognitive robotic system that integrates concurrent processes and interacting subsystems, performing according to different temporal properties and implementing the cognitive capabilities of interpretation, decision-making, learning, and communication [42].
- A cognitive system interpreting video sequences, and generating a natural language description of the observed environment employing situated graph trees [49].

4.6.2 Emergent Systems Approaches

These approaches employ self-organization processes for continually reconstituting the system itself in real-time such that to maintain its operational features via moderation of mutual system-environment interaction and co-determination. In emergent approaches, perception deals with the acquisition of sensory data in order to assure effective action [50]. In contrast to the cognivist approaches which use self-contained abstract models that are typically disembodied, and the physical instantiation of the system is not involved in the cognition model, emergent approaches are inherently embodied and the physical instantiation plays a significant role in cognition. Emergent systems involve three kinds of systems [43]:

- Connectionist systems
- Dynamic systems
- Enactive systems

They are based on the philosophical view that cognition is an emergent, dynamic and self-organizing process, in contrast with cognivist approaches that see cognition as symbolic, rational, encapsulated, structured, and algorithmic [51].

- *Connectionist systems* are massive parallel processing systems of non symbolic distributed activation patterns and involve the neural networks or neural computing systems [52, 53]. Neural networks used over the years for cognitive systems with various capabilities include multi-layer perceptrons (MLP), radial basis functions (RBF) neural networks, Hopfield neural network (HNN), self-organizing maps (SOM), Kohonen neural networks (KNN), and Grossberg's adaptive resonant theory (ART) networks.

- *Dynamic models* describe the perception-action cognitive processes and self-organize into metastable patterns of behavior [54]. They have very large numbers of degrees-of-freedom, and are open nonlinear non-equilibrium dissipative systems, i.e., they cannot maintain structure and behavior and dissipate (diffuse) energy. The nonlinearity assures a complex behavior, and a non-uniform dissipation of energy. The dynamic system approach is based on the philosophical view that dynamic models can directly provide many of the natural cognition characteristics such as adaptability, multi-stability, learning, intentionality, pattern formation, and recognition [55].

- *Enactive systems* are based on the philosophical view that cognition is a process by which the aspects that are required for the continuous existence of a cognitive agent are brought out or enacted, and co-determined by the agent during its interaction with the environment where it lies. This means that nothing is given *a-priori* and so there is no need for symbolic representations. Enactive systems are *autopietic*[1] (self-produced), i.e., they emerge as coherent systemic

---

[1]From the Greek "αυτό" (auto) and "ποιώ" (pio=make, produce).

although enaction exhibits different levels of self-production (cellular level, self-development, and structural adaptation through perturbation of their own organizational processes). Enactive systems identify regularities as a result of co-determination of the cognitive activities themselves so as to preserve system integrity, and operate in real-time. Enactive systems are rooted in the works of computational biologists such as [56].

Cognivist and emergent paradigms of artificial cognitive systems analysis and design can be distinguished using the following distinct *("cognitive vs. emergent")* criteria [43]:

- *Computational operation* (syntactic vs. self-organization).
- *Representational framework* (symbol patterns vs. skill construction).
- *Time constraints* (not entrained vs. synchronous real-time entrainment).
- *Inter-agent epistemology* (agent-independent vs. agent-dependent).
- *Embodiment* (not implied vs. cognition implied embodiment).
- *Perception* (abstract symbolic vs. response to perturbation).
- *Action* (causal vs. perturbation-based action).
- *Anticipation* (procedural vs. self-effected traverse).
- *Adaptation* (learning new knowledge vs. developing new dynamics).
- *Motivation* (resolve impasse vs. space interaction increase).
- *Relevance of autonomy* (not necessarily implied vs. cognition implied autonomy).

4.6.3 Hybrid Approaches

Cognivism was over the years and is still the predominant approach to cognition, but now seems not to keep its stronger interpretations. Both the *cognivist* (symbolic) and *emergent* (connectionist, dynamic, enactive systems) approaches fail in some aspects and succeed in others. Very often, when the cognivist approaches fail, the emergent approaches succeed and vice versa. Therefore, several hybrid cognivist-emergent systems that combine the best features of both were developed e.g., [57–59].

In [57] a hybrid cognitive robot control system is developed which integrates production systems, semantic networks, machine learning, and *subsymbolic processing* (**SS-RICS** system) to perform real-time mobile cognitive robot control. This system features high-order understanding and interaction with the dynamic world in which the robot lives.

In [58] a *"learning intelligent distribution agent"* (**LIDA**) system is proposed which, in addition to being a computational paradigm, it is a working model of human cognition. This human cognition model integrates neuro-computing and cognitive aspects, and possesses multiple learning capabilities during a developmental period. LIDA will be further discussed in Chapter 9.

In [59] a neural network-based perception-action component (based on exploratory learning) and a symbolic component (based on concepts and invariant descriptions) are combined to build a hybrid cognitive vision system.

Over the years many constructive criticisms on all approaches were provided. For example according to [60] cognivist systems suffer from three problems (the symbol grounding problem, the frame problem, and the combinatorial problem). Dynamic system theory has actually provided general modeling frameworks rather than cognitive systems, and has so far been used as an analysis tool than as a tool for the synthesis and design of cognitive systems. Hybrid systems can offer the best features of both worlds, but it is not clear how cognivist and enacting approaches should be combined for best behavior.

4.7 Cognitive Robot Design Requirements

We close our technical discussion with a set of requirements for the development of cognitive robots. These are the following:

- *Embodiment* (rich array of physical sensory and motor interface, morphology compatible with the model of cognition).
- *Perception* (attention on the goal of action, perception of objects, discrimination and addition of small number of objects, ability to learn hierarchical representations, attraction to people's faces, sounds, motion and features).
- *Action* (early movements with small number of degrees of freedom, navigation based on dynamic ego-centric path integration).
- *Adaptation and self-modification* (partial learning of affordances, autonomous generative model construction, transient and generalized episodic memories of past experiences, modification of both the structure and organization of the system).
- *Motivation* (explorative motives).
- *Autonomy* (preservation of homeostasis processes, minimal set of innate behaviors for exploration and survival, encoding of space in motor and goal specific manner).

The basic requirements for child-like robot development include the following:

- *Development of looking* (ability of the robot to redirect its gaze and move its eyes and head towards the object of interest, a pursuit model that enables the robot to fixate on a static or moving object dynamically tracking the object, and a model for predicting when and where the object reappears).
- *Development of reaching* (ability to reach towards a visual target first with a hand and then with its entire body, ability to adapt its reach towards the moving target and to learn when reaching is not possible).
- *Development of grasping* (ability to reach static or moving objects, to perform grasp closure during approach, and to efficiently match the grasp pose to a symmetry axis of the object).
- *Development of learning* (capability to learn by imitation through people tracking, face recognition, and action recognition).

A cognitive system architecture that satisfies many of the above robot-development requirements is presented in [61]. Figures 4.4–4.8 show snapshots of the developmental stages of looking, reaching, and grasping of infants and mental robots.

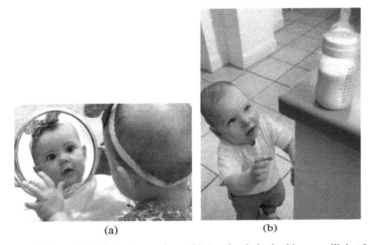

(a) (b)

Figure 4.4 (a) Baby looking at mirror. (b) Another baby looking at milk bottle.

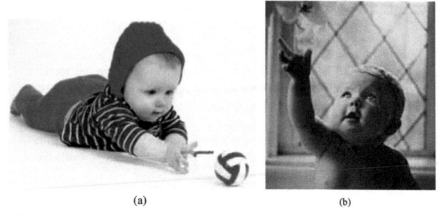

(a)   (b)

Figure 4.5   (a) A baby trying to reach a ball. (b) A baby trying to reach an orchid flower.

Source: (a) http://i.livescience.com/images/i/000/023/870/iFF/cute-baby-ball.jpg?1327441643
(b) http://cache2.asset-cache.net/gc/137388911-baby-girl-reaching-out-to-grab-orchid-flower-gettyimag
es.jpg?v=1&c=IWSAsset&k=2&d=Ohxg6fgSOnrhlpTTKNZ4w86sNRUWk2CqalH7BqxO868%3d

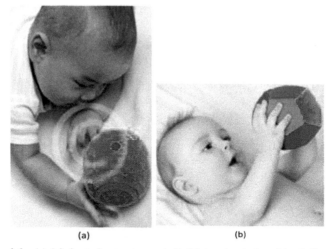

(a)   (b)

Figure 4.6   (a) A baby trying to grasp a ball. (b) Another baby with a ball grasped.

Source: (a) http://cdn.c.photoshelter.com/img-get/I0000FeONnDgbpW8/s/700/700/0508-058.jpg
(b) http://www.whenmybaby.com/images/babystats/baby_grasping_object.jpg

Figure 4.7 MIT social robot Leo looks at an object presented to it by a human.

Source: http://i.huffpost.com/gen/834593/original.jpg

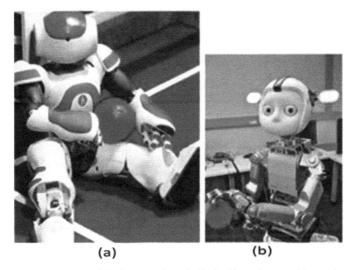

Figure 4.8 (a) Nao humanoid robot grasping a ball. (b) Simon humanoid grasping an object.

Source: (a) http://www.conscious-robots.com/images/stories/robots/aldebaran_nao.jpg
(b) http://www.engineersgarage.com/sites/default/files/imagecache/Original/wysiwyg_imageupload/1/
Simon_Meka-Robot.jpg

Figure 4.9 (a, b) shows two instances of baby and infant developmental learning.

Figure 4.10 shows two instances of robot developmental learning to perform particular tasks.

(a)   (b)

Figure 4.9   (a) Baby developmental learning to assemble cubes and cylinders. (b) Another instance of infant developmental learning.

Source: (a) http://www.child-development-guide.com/image-files/child-development-toys.jpg

(b) http://www.child-development-guide.com/image-files/child-development-and-learning.jpg

Figure 4.10   Two instances of robot developmental learning.

Source: (a) http://www.scientificamerican.com/media/inline/robot-learning_1.jpg

(b) http://www.sugano.mech.waseda.ac.jp/design/project/learning/img/2012030100.jpg

4.8 Concluding Remarks

In this chapter we have considered the fundamental aspects of human cognition and robot cognition. Important philosophical and practical questions in cognitive robotics are "whether a robot can learn as a child", and "whether the body has an impact on how we perceive and think". The first question is the subject of *"developmental robotics"*, and the second is considered in the framework of *"embodied cognitive robotics"*. Specifically, the chapter has addressed the ontological questions related to human cognition and robot cognition (cognitive robotics), and questions related to "philosophy *in* cognitive science", "philosophy *of* cognitive science", and "philosophy *of* cognitive robotics". The chapter has also presented some technological elements of cognitive robots based on the emergent systems approach, the cognivist approach, and the hybrid (cognivist-emergent) approach. Finally, the chapter has outlined a set of basic requirements for the design of cognitive robots. An important open research question in developmental robotics is what exactly "phylogenetic configuration" is needed for the autonomous development of mental capabilities (perception, action, adaptation, embodiment, anticipation, motivation, and autonomy) that enable the robot's ontogenetic development over its life time.

References

[1] J.R. Anderson, The Architecture of Cognition, Cambridge, MA, U.S.A.: Harvard University Press, 1983.
[2] U. Neisser, Cognitive Psychology, New York, NY, U.S.A.: Appleton-Century-Crofts, 1967.
[3] V. Eckardt, What is Cognitive Science? Cambridge, MA, U.S.A.: MIT Press, 1993.
[4] J. Piaget, The Grasp of Consciousness: Action and Concept in the Young Child, Cambridge, MA, U.S.A.: Harvard University Press, 1976.
[5] J. Locke, An Essay Concerning Human Understanding, New York, NY, U.S.A.: Oxford University Press, 1690.
[6] J.B. Watson, Behaviorism, New York, NY, U.S.A.: Norton, 1925.
[7] B.F. Skinner, Verbal Behavior, New York, NY, U.S.A.: Appleton-Century-Crofts, 1957.
[8] W. van Orman Quine, Word and Object, Cambridge, MA, U.S.A.: MIT Press, 1960.

[9] E.C. Tolman, Cognitive maps in rats and men, Psychological Review, 55, 1948, 189–208.

[10] E.C. Tolman, Collected Papers in Psychology, Berkley, CA, U.S.A.: University of California Press, 1951.

[11] N. Chomsky, Syntactic Structures, Paris, France: Mouton, 1957.

[12] N. Chomsky, Review of Skinner's Verbal Behavior, Language, 35, 1959, 26–58.

[13] A Turing, On computable numbers, Proceedings London Mathematical Society (2nd Series), 42, 230–265, 1936.

[14] A. Turing, Computing machinery and intelligence, Mind, 59, 1950, 433–450.

[15] W. McCulloch and W. Pitts, A logical calculus of the ideas immanent in neurons activity, Bulletin of Mathematical Biophysics, 5, 1943, 115–133.

[16] A. Newell and H. Simon, Computer Science as Empirical Inquiry: Symbols and Search, Communications of the ACM, 19(3), 1976, 113–126.

[17] A. Newell and H. Simon, Human Problem Solving, Englewood Cliffs, NJ, U.S.A.: Prentice Hall, 1972.

[18] D. Hebb, The Organization of Behavior: A Neuropsychological Theory, New York, NY, U.S.A.: John Wiley, 1949.

[19] G. Miller, Language and Communication, New York, NY, U.S.A., McGraw-Hill, 1951.

[20] G. Miller, The cognitive revolution: A historical perspective, Trends in Cognitive Sciences, 7(3), 2003, 141–144.

[21] J. Freedman, Robots Through History: Robotics, New York, NY, U.S.A.: Rosen Central, 2011.

[22] R. Brooks, Integrated Systems based on behaviors, SIGART Bulletin, 2(4), 1991, 46–50.

[23] R.A. Brooks, Intelligence without representation, Artificial Intelligence, 47, 1991, 139–159.

[24] R.C. Arkin, Motor-Schema based mobile robot navigation, International J. of Robotics Research, 8(4), 1989, 92–112.

[25] R.C. Arkin, Behavior-Based Robotics, Cambridge, MA, U.S.A.: MIT Press, 1998.

[26] C.F.B. Rooney, R.P.S. O'Donghue, B.R. Duffy, G.M.P. O'Hare, and R.W. Collier, The social robot architecture: Towards sociality in a real world domain, Proceedings Intern. Conf. Toward Intelligent Mobile Robots, Bristol, 1999.

[27] B.R. Duffy and G. Joue, Intelligent robots: The question of embodiment, Proceedings Intern. Conf. Brain-Machine, Ankara, Turkey, December 2000.

[28] J.H. Connel, SSS: A hybrid architecture applied to robot navigation, Proceedings IEEE Conference on Robotics and Automation (ICRA'92), 1992, 2719–2724.

[29] H.L. Dreyfus, Being-In-The-World: A Commentary on Heidegger's Being and Time, Cambridge, MA, U.S.A.: MIT Press, 1991.

[30] T. Smithers, On quantitative performance measures of robot behavior, J. of Robotics and Autonomous Systems, 15, 1995, 107–133.

[31] B.C. Smith, Situated Embeddedness, In: R.A. Wilson, and F.C. Keil (eds.), The MIT Encyclopedia of the Cognitive Sciences, Cambridge, MA, U.S.A.: MIT Press, 1999, 769–770.

[32] A. Brook, Introduction: Philosophy in and philosophy of cognitive science.

[33] C.L. Breazeal, Designing Sociable Robots, Cambridge, MA, U.S.A.: MIT Press, 2002.

[34] B. Adams, C. Breazel, R. Brooks, P. Fitz Patrick, and B. Scassellati, Humanoid robots: A new kind of tool, IEEE Intelligent Systems, 15(4), 2000, 25–31.

[35] A. Billlard and K. Dautenhahn, Experiments in learning by imitation: Grounding and use of communication in robotic agents, Adaptive Behavior 7(3–4), 2000, 415–438.

[36] S. Franklin, Artificial Minds, Cambridge, MA, U.S.A.: MIT Press, 1995.

[37] R. Preifer, Robots as cognitive tools, In: B. Gorayska and J.L. Mey (eds.), Cognition and Technology: Co-Existence, Convergence and Co-Evolution, Amsterdam, The Netherlands: John Benjamins, 2004, 109–126.

[38] A. Loula, R. Gudwin, and J. Queiroz (eds.), Artificial Cognition Systems, IDEA Group. Inc., 2006.

[39] R. Alami, R. Chatila, A. Clodic, S. Fleury, et al., Towards human aware cognitive robots, Proceedings 5th AAAI Intern. Cognitive Robotics Workshop (AAAI-06), Boston, U.S.A., Sept., 2006.

[40] N.L. Cassimatis, J.G. Trafton, M.D. Bugajka, and A.C. Schultz, Integrating cognition, perception, and action through mental simulation in robots, Robotics and Autonomous Systems, 49, 2004, 13–23.

[41] R.A. Brooks, and L.A. Stein, Building brains for bodies, Autonomous Robots, 1, 1994, 7–25.

[42] M. Scheutz, J. Kramer, C. Middendorff, P. Schermerhorn, M. Heilman, D. Anderson, and P. Bui, Toward affective cognitive robots for human robot interaction, Proceedings AAAI-05 Conf. on Artificial Intelligence, Pittsburgh, Penn., U.S.A., 2005, 1737–1738.

[43] D. Nernon, G. Metta, and G. Sandini, A survey of artificial cognitive systems: Implications for the autonomous development of mental capabilities in computational agents, IEEE Trans on Evolutionary Computation, 1(2), 2007, 151–177.

[44] Z.W. Pylyshyn, Computation and Cognition, Bradford Books Series, Cambridge, MA, U.S.A.: MIT Press, 1984.

[45] R.N. Shepard and S. Hurwitz, Upward direction, mental rotation, and discrimination of left and right turns in maps, Cognition, 18, 1984, 161–193.

[46] A. Newell, Unified Theories of Cognition, Cambridge, MA, U.S.A.: Harvard University Press, 1990.

[47] B. Neumann and R. Möler, On scene interpretation with description logics, In: H.I. Christensen and H.-H. Nagel (eds.), Cognitive Vision Systems: Sampling the Spectrum of Approaches, Berlin, Germany: Springer, 2005, 235–26.

[48] J.L. Crowley, Things that see: Context-aware multi-modal interaction, Special Issue on Cognitive Computer Vision: KI-ZK Intelligenz, April 2005.

[49] H.-H. Nagel, Steps toward a cognitive vision system, AI Magazine, 25(2), 2004, 31–50.

[50] W.D. Christensen, and C.A. Hooker, An interactive-constructivist approach to intelligence: Self-directed anticipative learning, Philosophy and Psychology, 13(1), 2000, 5–45.

[51] D. Vernon, Cognitive vision: The case for embodied perception, Image and Vision Computing, 2006, 1–14.

[52] S. Haykin, Neural Networks: A Comprehensive Foundation, Upper Saddle River, NJ., U.S.A.: Prentice Hall, 1999.

[53] M.A. Arbib (ed.), The Handbook of Brain Theory and Neural Networks, Cambridge, MA, U.S.A., MIT Press, 2002.

[54] E. Thelen, and L.B. Smith, A dynamic system approach to the development of cognition and action, Bradford Book Series in Cognitive Psychology, Cambridge, MA, U.S.A.: MIT Press, 1994.

[55] J.S. Kelso Dynamic Patterns: The Self-Organization of Brain and Behavior, Cambridge, MA, U.S.A.: MIT Press, 1995.

[56] H. Maturama, and F. Varela, The Tree of Knowledge: The Biological Roots of Human Understanding, London, U.K.: New Science Library, 1987.

[57] E. Avery, T. Kelley, and D. Davani, Using cognitive architectures to improve control: Integrating production systems, semantic networks, and sub-symbolic processing, system, 77, 1990; Also: Proceedings of 15th Annual Conf. on Behavioral Representation in Modeling and Simulation (BRIMS), Baltimore, MDD, 2000, 15–18.

[58] U. Ramamarthy, B.J. Baars, S.K. D'Mello, and S. Franklin, LIDA: A working model of cognition, Proceedings of Intern. Conf. on Cognitive Modeling, 2006, 244–249.

[59] G. Granlund, A cognitive vision architecture integrating neural network with symbolic processing, Special Issue: Cognitive Computer Vision: KI-Zeitschrift Künstiche Intelligenz, April 2005.

[60] W.D. Christensen, and C.A. Hooker, Representation and the meaning of life, In: Representation of Mind: New Approaches to Mental Representation, Sydney, Australia, University of Sydney, 2000.

[61] M. Rudinac, Exploration and Learning for Cognitive Robots, Doctoral Thesis, Technical University of Delft, Netherlands, 2013.

5

Robot Intelligence

---

*Intelligence is the computational part of
the ability to achieve goals in the world.
Varying kinds of intelligence occur in people,
many animals, and some machines.*
John McCarthy

*The ability to improve behavior through
learning is the hallmark of intelligence,
and thus the ultimate challenge of AI and robotics*
Maja J. Mataric

*The real problem is not whether
machines think but whether men do.*
B.F. Skinner

5.1 Introduction

Intelligence is a multifaceted feature of humans which was one of the study
subjects of psychology right from its beginning. Individual persons have
different abilities to comprehend complex concepts, adapt efficiently to the
environment, learn through experience, imitation or self-correction, overcome
obstacles, avoid risks and generalize various types of reasoning. These individual differences are never consistent because the intellectual performance
of the same person varies, according to the particular situation or area of
activity, as a result of different criteria. Over the years a great effort was
put by psychologists and other thinkers to clarify and organize the complex
phenomena of human intelligence. However, so far no single definition (or
theory) of intelligence, universally acceptable, exists. Human intelligence

99

(its nature, its mechanisms, etc.) is still a matter of debate. Psychologists have proposed several criteria (tests) for measuring intelligence. But many current theoretical scientists argue that there are many issues and features of intelligence that cannot be captured by standard psychometric tests, while others point out the role that culture and society play in establishing several conceptions of intelligence and influencing the acquisition of intellectual skills.

One of the most fundamental issues in the area of machine intelligence is whether the machines will ever be able to think like humans. *Alan Turing* in 1950 tried to give an answer to this question and proposed a test that can be used to judge whether machines are genuinely thinking. This is well-known as the *Turing test* for machine intelligence. Over the years the Turing test has attracted extensive criticism based on the view that a computer can only do what we program it to do (which means that it only has a fixed pattern of behavior). However, many artificial intelligent workers claim that passing the Turing test is a good reason to judge that a computer has really thoughts and other mental states, while others (the behaviorists) accept that passing the Turing test suffices for being intelligent. The real problem here is that no computer has yet passed this test in its generality and unconstrained form, and many scientists believe that none ever will (see, for example, the debate of Mitchell Kapor and Kurzweil on Longbets.org site: www.longbets.org/1/). Actually, there are many examples of computers that passed restricted forms of Turing test and many other which show that machine intelligence, no matter how it is defined and interpreted, is not unlimited.

The purpose of this chapter is to study some fundamental philosophical aspects of human intelligence and machine intelligence.

Specifically the chapter:

- Examines the ontological issues of human intelligence "what is", "what are its forms", and "how it is measured".
- Discusses the same ontological aspects of machine-robot intelligence.
- Provides the basic milestones of machine intelligence.
- Summarizes the philosophical replies of eminent machine intelligence scientists about the ontological and phenomenological questions of machine-robot intelligence.
- Reviews the Turing machine intelligence test.
- Outlines some other representative machine intelligence tests (text compression test, psychometric-like test, universal test, symbolic machine intelligence test, and multi-dimensional test).

5.2 What Is Intelligence?

Human intelligence is perhaps the most controversial concept of study in psychology. It has been defined indirectly in several ways in terms of its features and capabilities [1–3]. Etymologically, the word intelligence comes from the Latin word *"intellegere"* (to understand, to perceive, to recognize or to realize). The first part of the word is derived from the prefix *"inter"* (which means *"between"*) and the second part comes from the word *"legere"* (meaning to select, to choose, and to gather). The combination of these two words can be interpreted as the capacity to establish abstract links between details that have not any obvious relationship.

Some of the definitions that are available in the dictionaries are the following:

- *Free Dictionary:* Intelligence is the capacity to acquire and apply knowledge and reason, the faculty of thought, understanding and act.
- *Cambridge Dictionaries:* Intelligence (cleverness) is the ability to understand and learn, and make judgments or have opinions that are based on reason.
- *Merriam-Webster:* Intelligence is the ability to learn or understand or to deal with new trying situations; also the skilled use of reason.

Four other definitions given by particular thinkers are:

- *Dearborn:* Intelligence is the capacity to learn or to profit by experience.
- *Wechsler:* Intelligence is a global concept which involves a person's ability to think rationally, to act purposefully, and to deal effectively with the environment.
- *Bingham:* Intelligence means the ability of an organism to solve new problems.
- *Legg and Hutter:* Intelligence measures an agent's ability to achieve goals in a wide range of environments.

Psychology defines the intelligence as a general term encompassing several mental abilities, namely the abilities of reasoning, perceiving relationships and analogies, learning, memorizing, using what is learned, calculating, adapting to new conditions, handling the environment, etc.

Robert Sternberg (1985) has proposed his "triarchic *theory*" which includes three fundamental aspects or abilities of intelligence namely [4]:

- *Analytic abilities* (which enable one to analyze, evaluate, judge, compare, and contrast).

- *Creative abilities* (enabling a person to create, invent, imagine or discover).
- *Practical abilities* (enabling a person to apply, utilize, implement, build, and act).

His studies showed that intelligence can be defined as "the comparative level of performance of a human in reaching his own objectives". A human with greater intelligence, in the same situation, reaches his/her goals more frequently or with relatively higher speed. He argues that successfully intelligent persons discern their strengths and weaknesses, and then figure out how to capitalize on their strengths and to compensate or heal their weaknesses. Sternberg and his co-workers have developed measures of *"tacit knowledge"* in various domains such as business management, etc. Other researchers have demonstrated the relative independence of academic and practical intelligence.

An alternative 3-level definition of intelligence in terms of its capabilities which combines the above definitions is the following.

"Intelligence includes three hierarchical capabilities, namely:

- *Learning ability* (ability to learn from experience, supervised training or educations in all possible ways, formal or informal).
- *Problem formulation ability* (ability to identify problems and vague situations and formulate the problem concerned in a clear way).
- *Problem solving ability* (ability to solve problems, achieve goals, and design products)".

It is widely accepted that intelligence is inherited, but can also be associated with the environment. Many studies have demonstrated that heredity is an important factor in determining intelligence, but others suggest that the environment is a crucial factor in determining the extent of its expression. The level of one's intelligence determines how well he/she faces changes of the environment. George Santaya says: *"Intelligence is quickness in seeing things as they are".*

Many psychologists have argued that intelligence can be quantified mainly through testing. *Alfred Binnet* and *Theodore Simon* (1905) devised a system for testing intelligence with scoring based on average mental levels for various age groups [5]. They observed that the intelligence level of children is directly related to age. For each age they designed lists of questions and abandoned those which have been erroneously answered by more than 25% of the "normal" children sample. In this way they obtained a set of questions appropriate and adapted to each age, i.e., such as to mirror the average knowledge of the children of that age. The *mental age* (**MA**) is determined as follows. If an individual child answers the questions of age x but fails to answer the questions of age x + 1, then the child has mental age MA = x.

The quotient of the mental age **MA** thus obtained and the actual *chronological age* (**CA**) of the child, called *Intelligence Quotient* (**IQ**), gives a decimal number which in percent form is expressed as

$$IQ = 100 \text{x} (MA/CA)$$

The experiment has shown that the statistical distribution of IQ has a bell form (almost Gaussian) which is known as the *IQ Bell Curve* [7], and has a standard deviation $\sigma = 51$ about the average IQ $= 100$. Most people cluster around the average IQ $= 100$. Few are very bright or very dull. It was verified that approximately 65% of the people have scores between one standard deviation ($85 < $ IQ $ < 115$) and about 95% of the population have scores between two standard deviations ($70 < $ IQ $ < 130$). Nowadays, the above interpretation of IQ is no more widely used. Many of the most popular intelligence tests are not intended to measure intelligence itself, but some closely related construct (**SSAT** for preparatory school, **SAT** or **CAT** for college, **GRE** for graduate school, **MOAT** for medical school, **LSAT** for law school, and **GMAT** for business school). Tests are used for several purposes such as selection, diagnosis, and evaluation.

Individuals rarely perform equally well on all different aspects included in an intelligence test. For example, somebody may perform relatively better on spatial than on verbal items, while somebody else may do the opposite. However, it was verified that subtests concerning different skills tend to be positively correlated, i.e., people who do well on one subtest are likely to score above average on others as well. This has led to the concept of a more general intelligence factor, the well known *factor* **g**, which indicates what all tests have in common [6]. Some psychologists suggest to examine more specific group factors such as counting ability, verbal comprehension ability, memory, etc. Of course, over the years there have been many debates about the utility of IQ and g. Some psychologists criticize the entire psychometric methodology's tests, while others believe that this methodology is firmly justified and established [7]. Those who argue against the test scores, say that measuring intelligence using only test scores, many important aspects of mental ability are ignored. Actually, intelligence is completely dynamic (i.e., it is evolving, always changing, progressing, and never ending). When intelligence is quantified, it becomes a static feature, and so scores can only be regarded as *instantiations (snapshots) of intelligence.*

Some important questions about human intelligence that are still under investigation are the following [8]:

- What are the current dominant features and conceptualizations of intelligence?
- Are intelligence tests useful? What is their meaning? What do they predict, and how efficiently?
- Why do people have different intelligence as measured by intelligence tests? Do the differences depend on genetic, ethnic, educational aspects, etc?
- Are there any important aspects of intelligence still unresolved? If yes why?

These issues were the subject of many discussions over the years, especially after the presentation of the controversial *Bell Curve* concept [7]. Intelligence can be distinguished in several types, some of which are the following [8] (Figure 5.1):

- *Existential intelligence* (Sensitivity and capability to face deep questions about human existence, the meaning and purpose of life, etc.).
- *Naturalistic intelligence* (capability to discriminate among living creatures as well as other features of the world, such as rock configurations, clouds, etc.).
- *Interpersonal intelligence* (capability to understand and interact effectively with others with verbal and nonverbal communication).
- *Intrapersonal intelligence* (capability to understand oneself and one's thoughts and feelings, and use them to plan and direction one's life).
- *Logical-mathematical Intelligence* (capacity to calculate, quantify and study propositions and hypotheses, and perform complex mathematical operations including abstract symbolic logic). This intelligence is usually developed in mathematicians, scientists and engineers.
- *Verbal-linguistic intelligence* (ability to think in words and to use language to express and employ complex meanings).
- *Bodily-kinesthetic intelligence* (capability to manipulate objects and use several physical skills requiring mind-body union).
- *Visual-spatial intelligence* (capability to think in three dimensions including mental imagery, spatial reasoning, image manipulation, graphic and artistic skills, etc.).
- *Musical-rhythmic intelligence* (capacity to discern pitch, rhythm, timbre and tone, capability to recognize, create, reproduce, and reflect on music demonstrated by composers, musicians, conductors, etc.).

Figure 5.1 Howard Gardner's multiple intelligences.

Source: www.intelligence1weebly.com/uploads/2/6/0/9/26090838/9177282_orig.png

It is not expected that any individual person shows the same performance in all these aspects of intelligence, although he/she may be improved through education, training, and practice.

5.3 What Is Machine-Robot Intelligence?

Like human intelligence, machine and robot intelligence is difficult (if not impossible) to be defined uniquely and completely, and its definition is still a great deal of debate. The creation and study of machines that are *"intelligent"* is the central concern of the *artificial intelligence* (**AI**) field. The term "artificial intelligence" was coined by John McCarthy in 1956 (at the Dartmouth Conference), but the term machine intelligence was coined earlier (1950) by Alan Turing in his seminal paper "Computing Machinery and Intelligence" [9].

The theoretical foundations of AI include:

- *Philosophy* (Logic, reasoning methods, learning, language, rationality).
- *Psychology* (Cognition, perception, motor control, experimental methods).
- *Mathematics* (Mathematical representation, proof of algorithms, computation, probability).
- *Neuroscience* (Physical substrate for mental ability).
- *Computer engineering* (Building powerful and fast computers).
- *Linguistics* (Knowledge representation, grammars).
- *Control systems* (Feedback, optimal, adaptive and predictive control algorithms).
- *Economics* (Decision theory, utility theory, game theory, and economic models).

A general simple definition of AI could be the following:

"Artificial intelligence is the study of systems that behave and act such that to any observer would appear to be intelligent".

Machine intelligence has been defined over the years indirectly through the definition of *intelligent machine*. Three definitions of an intelligent machine available in the literature are the following [10–13]:

Definition 1 *(McGraw Dictionary)* Intelligent machine is a machine that can accomplish its specific task in the presence of uncertainty and variability in its environment.

Definition 2 An intelligent machine is a system (computer, robot, etc.) that emulates some aspects of intelligence exhibited by nature. These include learning, adaptability, reasoning, problem solving, robustness, and improving efficiency.

Definition 3 Intelligent machine is an intelligent agent, where *"agent"* is an entity that perceives and acts in the environment. The success of an agent is measured by a *"performance index"*, and the agent is intelligent if its actions optimize the expected value of the performance index on the basis of past experience and knowledge. This is the so-called *"agent-based definition* of an intelligent machine".

As argued in [11], AI is evolved in two principal directions:

- *Humanistic AI* (**HAI**) which is concerned with machines that think and act like humans.
- *Rationalistic AI* (**RAI**) which is concerned with machines that think and act rationally and are built on the basis of understanding intelligent human behavior.

HAI attempts to create and study machines that perform functions which need intelligence when performed by people. It is the field that studies how to make computers do things for which people are, at the moment, better [14]. This would mean that HAI is somehow *ephemeral* because it refers to the current state of art of computer science, but actually the rate at which this meaning of HAI might change is nowhere near as great as one might think.

RAI attempts to explain and emulate intelligent behavior in terms of computational processes. It is the field of study which is concerned with automating intelligent behavior. Today, intelligent machines seem to do more with rationalistic than humanistic AI.

Thinking humanly involves the cognitive modeling approach which studies the internal actual processes of the human mind. *Acting humanly* takes the *Turing Test* approach. The Turing test suggests that if a machine can answer any question put to it, using the same words that an ordinary person would, this machines may be called intelligent. *Thinking rationally* is founded on the laws of thought (i.e., to Logic). Finally, *acting rationally* takes the agent approach which characterizes an agent as intelligent if it can observe the state of its environment, and possesses a set of actions via which it changes one state to another so as to optimize its performance at will.

5.4 Milestones of Machine Intelligence

The history of intelligent machines and artificial intelligence is obviously overlapped with the history of cognition and cognitive robotics discussed in Section 4.4. Here some milestones specific to artificial intelligence will be briefly summarized [15, 16]:

1642 *Pascal* creates the first mechanical digital calculating machine.
1673 *Leibniz* improves Pascal's machine to perform multiplication and division (the so-called *Step Reckoner*).
1801 *Joseph-Marie Jacquard* invents the *"Jacquard loom"*, the first programmable machine, with instructions on punched cards.

1832 *Charles Babbage* and *Lady Lovelace* (Ada Byron) design a programmable mechanical calculating machine, called the *"Analytical Engine"*.

1854 *George Boole* in his book "The Laws of Thought" presents the binary algebra that represents some thinking laws.
Bertrand Russel and *Alfred North Whitehead* publish "Principia Mathematica" which revolutionalized formal logic and led philosophy into logical analysis of knowledge.

1950 *A.M. Turing* publishes "Computing Machinery and Intelligence" and introduces the *Turing Test* as a way of assigning intelligent behavior to machines. Also: *Claude Shannon* presents detailed analysis of chess playing search in his "Programming a Computer to Play Chess".

1956 This year is recognized as the formal starting year of artificial intelligence (**AI**), with *John McCarthy* coining the term *"artificial intelligence"* as the topic of the Dartmouth Conference (the first conference on this subject).

1957 *Newell, Shaw,* and *Simon* demonstrate an AI system for general problem solving *(GPS)*.

1958 *John McCarthy* develops the List Programming Language (LISP).

1964 *Danny Bobrow* shows that computers are able to understand natural language such that to solve algebra problems correctly.

1966 *Ross Quillian* demonstrates semantic nets, and *Donald Michie* organizes the first Machine Intelligence workshop at Edinburgh.

1967 *Edward Feigenbaum* (and colleagues) present the *DENTRAL program,* the first successful knowledge-based expert system for scientific reasoning.

1970 *Patrick Winston* shows how to implement learning of concepts from examples (the block words problem).
Bill Woods describes Augmented Transition Networks (**ATN**) for the representation of natural language understanding.

1972 *Alain Colmerauer* develops Prolog Programming language.

1975 *Marvin Minsky* publishes his first work on *Frames* as a representation of knowledge.

1979 *Ted Shortliffe* presents the **MYCIN** system, the first expert system in the medical area.

1980s IBM starts the IBM PC Compatible (1981). Lisp machines are commercialized, **HEARSAY II** speech understanding system is implemented using *Rick Hayes-Roth's "blackboard model"*, the first expert systems shells are commercialized, and neural networks are widely used with the *back propagation* (**BP**) algorithm developed by Werbos in1974.

1990–2000 Major advances in AI subfields (machine learning, case based reasoning, uncertain reasoning, scheduling, multi-agent planning, natural language understanding, virtual reality, ontologies, etc.) and intelligent robots (Rod Brook's intelligent humanoid robot **COG**, NASA's path finder intelligent robot called **Sojourner**, robotic soccers, interactive socialized humanoids, zoomorphic, and mobile robots). On the theoretical side, important advances in intelligent robot control theory with generic architectures, implementations, and commercialization (knowledge-based control, computational intelligence-based control, Petri-net based control, vision-based control, etc.).

5.5 Philosophical Issues of Machine Intelligence

It has been a diachronic debate among philosophers and scientists on how far intelligent machines can go. Neither side of the contention was able so far to provide solid arguments of proving this way or the other.

Philosophical questions of concern about machine intelligence include:

- Will computers ever be able to genuine think at all?
- Are any fundamental differences between human thought and any thought that computers will be ever capable?
- Will computers ever really be able to have genuine thoughts, intelligence, self-consciousness, a real mental life, even if it might be different in some respects from our own?
- Will computers ever be able to solve any problem that a person would solve by thinking?

In the following, we summarize a set of representative answers to these questions (positive and negative) given over the years by eminent AI scientists, roboticists, and philosophers.

John McCarthy (1956) stated that: "every aspect of learning or any other feature of intelligence can be so precisely described that a machine can be made to simulate it" (Dartmouth Conference).

Allen Newell and Herbert Simon (1958) depicted the future of machine intelligence by saying: "There are now in the world machines that think, that learn, and that create. Moreover their ability to do these things is going to increase rapidly until (in a visible future) the range of problems they can handle will be co-extensive with the range to which the human mind has been applied" [17]. With their 1976 *"Physical Symbol System Hypothesis"* **(PSSH)** concept [18] they claimed that human thinking is a kind of symbol manipulation and machines can be intelligent. They defined a PSSH as a system that accepts physical patterns (symbols), combines them into structures (expressions), and manipulates them (using processes) to produce new expressions. The physical symbol system hypothesis has its origin to the philosophical ideas of *Hobbes* ("reasoning is nothing more than reckoning"), and *Hume* (perception can be reduced to "atomic impressions"). Specifically, PSSH states: "A physical symbol system has the necessary and sufficient means for general intelligent action".

Hubert Dreyfus (1972) argued that "computer simulation workers assume incorrectly that explicit rules can govern intellectual processes". One of his arguments was that computer programs are inherently *goal seeking,* and thus require the designer to know beforehand exactly what behavior is desired (as, e.g., in chess match) as opposed to a work of art. Contrary to this, humans are *value seeking,* which means that they don't always begin with an end goal in mind, but seek to bring implicit values to fruition, the fly, through engagement in a creative or analytical process" [19, 20].

Kurt Gödel (1931) proved that one can always create statements that a formal system, such as a symbol manipulation program, could not be able to prove. However, a human being can, with some reasoning, see the truth of these statements (called Gödel statements). The philosopher *John Lucas,* based on Gödel's arguments stated that human reason would always be superior to machines. In his own words: "Gödel's theorem seems to me to prove that mechanism is false, that is, that minds cannot be explained as machines" [21].

S. Russel and P. Norvig (1995) pointed out that after 20 years from Dreyfus' critique, "AI rules" were discovered that govern unconscious reasoning and skills. Neural networks, genetic algorithms, etc., are mainly attempted to capture human's unconscious capabilities of perception and attention. AI was shifted from symbol-based manipulation toward non-symbolic manipulation.

Daniel Crevier (1993) has argued that the reality has proved the correctness of some of Dreyfus' critical concerns about AI (although, had he stated them less aggressively, constructive suggestions might have been made much earlier) [22].

John Searle (1997), a philosopher at University of California, Berkeley, challenged the concept of machine intelligence and rejected the *strong AI's* claim that "the mind is just a computer program" [23]. His *"Chinese room argument"* is that "computers are immensely useful devices for simulating brain processes, but the simulation of mental states is no more a mental state than the simulation of an explosion is itself an explosion". John Searle has distinguished *strong AI* from *weak AI* which assumes that "physical symbol system can act intelligently". On this issue Norvig agrees that "Most AI workers take the *weak AI* hypothesis for granted, and don't care about the *strong AI* hypothesis [11].

Hans Moravec (1988), a roboticist, suggested that humans must give way to the new species of intelligent machines (robots). He says: "We should keep researching, and should proudly work to create robots that will supplant humans as Earth's superior species. Humans should just get out of the way of this self-imposed evaluation" [24].

Roger Penrose (1999) argued that "true intelligence cannot be presented without consciousness, and hence intelligence can never be produced by any algorithm that is executed on a computer", or in other words that "the conscious minds can find no home within our present-day scientific world" [25, 26].

Ray Kurzweil (1999), a computer scientist and futurist, takes further Moravec's position on robot species and argues that "the human brain presumably follows the laws of physics, so it must be a machine, albeit a very complex one. Is there an inherent difference between human thinking and machine thinking? Or, in other words: "Once computers are as complex as the human brain and can match the human brain in subtlety and complexity of thought, are we to consider them conscious? Computers will appear to have spiritual experiences, and people (those still using Carbon-based neurons) will believe them [27].

Gilder and Richards (2002) commented Kurzweil's utopia by saying: "Kurzweil's record as a technology prophet spurred interest in this more provocative prediction that within a few decades, computers will attain a level of intelligence and consciousness both qualitatively and quantitatively beyond human capacity" [28].

William Bainbridge (2004) referring to the possible impacts of robots on human's longevity said: "In principle, and perhaps in actuality, three or four decades from now it should be possible to transfer a human personality into a robot, thereby extending the person's lifetime by the durability of the machine" [29].

Douglas Hofstadler (1999), a mathematician and philosopher, on the question: "will emotions be explicitly programmed into a machine?" gave a firm negative answer: No! He says: "That is ridiculus. Any direct simulation of emotions cannot approach the complexity of human emotions, which raise indirectly from the organization of our minds. Programs or machines will acquire emotions in the same way: "as by products of their structure, of the way in which they are organized – not by direct programming" [30].

Jeff Hawkins (2004), a Silicon Valley computer architect, completely rejects the possibility of human-like computers. He says: "Can computers be intelligent? For decades scientists in the field of artificial intelligence have claimed that computers will be intelligent when they are powerful enough. I don't think so. Brains and computers do fundamentally different things" [31].

Jinchang Wang (2013) argues that "machine intelligence is not unlimited and that there exists an insurmountable barrier for digital computers and robots to achieve the full range of human intelligence" [32]. He bases this position by showing that "If a robot has a human sentience of life and death then it would cause a logical contradiction". To this end, he makes the implicit assumption that "we are not able to duplicate an existing person's *self* by any technology such as programming". Therefore, a robot cannot have the self-consciousness of any one existing in the world. If a person knew how to copy himself then he would lose the sentience of *anxiety of death* due to the existence of his "copies". Death of a robot can be avoided using a "copying program", but for the human, death is the *"sure end"*. Anxiety of death is a fact of human consciousness, which excludes coping. "Since computers are copyable, robots controlled by computers cannot have anxiety of death. Thus robot intelligence is not unlimited".

The philosophical interest in the computer revolution has so far focused on the as-yet unfulfilled promise of artificial intelligence. The two primary and tightly jointly-entangled questions of consideration were (and still are) whether a thinking machine is possible, and whether such a machine can provide a proper model of the human mind. Today, computer scientists and AI workers, while not abandoning their philosophical concerns, focus their attention more on the use of the up-to-date accomplishments in AI, combined with the networked personal computers, for designing AI systems of practical value in the modern society. Such systems include medical and service robots, automatic car driving, business forecasting, Internet searching, stock market, etc. Although emulating human thinking on a machine remains the principal goal of AI, it seems a goal not so easily achievable as people initially think. Until now artificial intelligence mimics only some aspects of human intelligence and cannot think truly autonomously.

Some other more specific philosophical questions of great concern in machine intelligence are:

- Can a machine be creative?
- Can a human' s self be copied?
- Can an existing person's mind be programmed?
- Can a machine be benevolent or harmful?
- Can a machine be self-aware?
- Can a machine be non copyable? If so how does it work?
- Can a machine have emotions? If so, are they different than human emotion?
- Can machine intelligence be quantified? If so, how?

In the next sections we will briefly discuss the last question. Inspired by the *intelligent quotient* (**IQ**) and the *factor g* developed by psychologists for the measurement of human intelligence, AI workers have devoted much effort to study the question of whether machine intelligence can be measured and how. This work has started with the famous *Turing Test* which is debated and extended or modified since the time of its presentation in 1950 [9].

5.6 The Turing Machine Intelligence Test

The Turing test works as follows (Figure 5.2). A human judge performs remote conversations (through teletype) with another human and a computer, and must guess which is the computer and which is the human. If the computer can very often fool the human judge into thinking it is human, that computer passes the test, and Turing claims that the computer is intelligent and can genuinely think.

Turing is only claiming that passing this test is a *sufficient* condition for the machine to be intelligent. He doesn't want to claim that being able to pass his test is a *necessary* condition for being intelligent. He only argues that the machines that can pass his test are intelligent. Also, Turing was not concerned about who is supposed to be judging these tests. But that's very important, because it is very easy to fool computer novices or other people into thinking that some program is really intelligent, even if the program is actually fully stupid. In the literature there are many example of this. But if we weaken Turing test by saying that passing the Test is a *real sign* that some program is intelligent, then it doesn't matter who is judging the Turing test. We can use

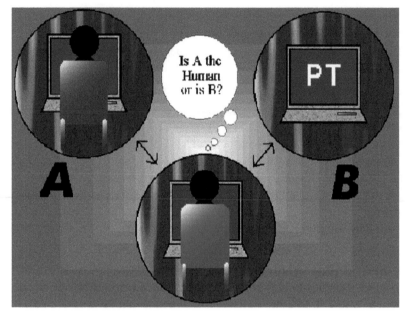

Figure 5.2   Pictorial illustration of the Turing test.

Source: https://huehueteotl.files.wordpress.com/2007/04/turing_test.gif

better judges than just non expert people. One straight forward objection to Turing test is that computers are not error-free as it is tacitly assumed. They can make mistakes (functional and arithmetic) like humans.

This simple and elegant Turing Test has received much criticism. For example Searle [23] states that passing the test is not sufficient to establish intelligence, because a machine might seem intelligent without having any "real intelligence", perhaps by using a very large repertory of answers to questions (something that might be practically impossible due to the vast question-answer table needed). Another difficulty of the Turing test is that present day AI systems are far away from being capable to pass an unrestricted Turing test. Actually, the Turing test is an example of the *behaviorist* view about the mind, which for the case of intelligence says that all there is to being intelligent is behaving, or being disposed to behave, in certain ways in response to specific types of stimulation (See Section 4.4.1). To overcome the above (and other) problems of the Turing test many more demanding versions of it have been proposed in the literature, e.g., the *Total Turing test,* the *Truly Total Turing test,* the *Inverted Turing test,* etc. [33–35]. The Turing test (originally conceived just as an imitation game) has inspired the answers to many philosophical questions about thinking machines.

On May 11th, 1997, the IBM computer named *Deep Blue* was able in chess playing to "outthink" the World Chess Champion, Garry Kasparov, with two wins and three draws. Later, the IBM supercomputer, named *Watson,* competed against previous champions on the television quiz show *Jeopardy*! Watson was able to defeat Brad Ratter, the biggest all-time money winner on Jeopardy!, and Ken Jennings, the record holder for the longest championship streak on Jeopardy!

The above are two examples that computers can defeat humans at specific tasks, but this is far from passing a real Turing test, or having the necessary proof that a machine can truly think like humans. What is true here is that, so far, computers are better than humans at two specific functions, namely:

- The ability to store and retrieve enormous quantities of data.
- The proficiency at calculation.

Humans also use memories to solve problems based on past experience, but they accumulate them via the process of learning. This ability of independently learning, is a prerequisite for exhibiting intelligence, and is not possessed by computers. The calculation superiority of computers is due to their capability to search and retrieve possible solutions and then compute the relevancy on the basis of available datasets in order to come up with a proper answer.

In a recent competition 7 June 2014 at the University of Reading (U.K.) the best five machines from around the world came together, and the winning supercomputer program named *"Eugene Goostman"* (developed at Saint Petersburg, Russia) managed to fool 10 judges, out of 30 competent software engineers, scientists, philosophers and journalists, into thinking there were talking to a 13-year old boy from Odessa, Ukraine. The creators of Eugene and the organizers of the competition proudly declared that this is a case of passing the Turing test, which so far no computer has managed to pass. But the scepticists rejected that Eugene has passed a true Turing test, with several arguments (see, e.g., www.bbc.com/new/technology-27762088).

5.7 Other Machine Intelligent Tests

5.7.1 General Issues

As seen in Section 5.2, human intelligence is tested (and quantified) by the psychometric indexes IQ and g which have attracted much criticism about their validity, generality, and static characteristics. The IQ can, at most, indicate the minute (static) performance, but does not provide a good score for a long period or for the future capacity of the individual. Binnet's IQ provides only a socially and culturally biased measure of the "already developed abilities", and just a small hint, if any, to the "potential intellectual abilities" of the person tested. Knowing the above difficulties of psychometric tests and the problems of the Turing test mentioned above, AI workers have developed, and are still developing several machine intelligence tests with various features, levels of success, and applicability. However none of these tests are real Turing tests.

The primary features that characterize the tests of machine intelligence, and are used to compare them, are the following [36]:

- *Validity:* This feature concerns the requirement that a test should capture intelligence and not some related property.
- *Range:* This refers to the range of intelligence for which a test is applicable (the range should be as a larger as possible from very low to extremely high intelligence).
- *Generality:* The more generality a test has the better. A test must be as more general as possible to cover everything, if possible (from a fly to a learning algorithm).
- *Dynamic:* Tests should be dynamic taking into account the possible improvement of the individual's intelligence via learning and adaptation.
- *Formal:* Tests should be defined precisely (perhaps by using mathematics).
- *Fundamental:* A test must be invariant to changes of technology and knowledge.
- *Objectiveness:* Subjective assessments (e.g., the opinion of human judges) should not be involved in the test.
- *Unbiased:* A test should not be biased in any respect (e.g., species, culture, etc.).
- *Fully defined:* All aspects of a test should be clearly defined, or at least the aspects not specified should be identified.
- *Universal:* A universal test (not merely anthropomorphic) is preferable. Universality is significant in order to ensure that the heterogeneity of machines is not a problem.
- *Practical:* Tests must be practical i.e., capable to be carried out quickly and automatically.
- *Test vs. definition:* A method proposed for testing machine may be more a test or more a mere definition or in between.

In practice (industry, robotics, business, telecommunications, etc.) it may be preferable to have a more active or efficient machine with smaller intelligence, than a more brilliant but lazy (or dormant) machine. Let MI be the *machine intelligence* and W the *working capacity* of the machine (e.g., its speed of calculation or speed of data transfer). Then, we define the combined index P as [37]:

$$P = MI \times W$$

In practice, machine with greater P is desirable. It is the product MI×W that matters rather than simply MI. A machine with very high W may provide a fine solution to a problem in much shorter time rather than a machine with greater MI.

Another machine intelligence quotient, useful in practice, is based on the unified definition of intelligence provided in [37], which states:

"Intelligence measures an agent's ability to achieve goals in a wide range of environments".

This concept of intelligence actually measures an agent's *power* to optimize the world according to its preferences. Therefore, it can be called the *optimization power* (**OP**) concept of intelligence. This view of intelligence is very useful in practice since it quantifies intelligence with externally measurable performance rather than the details of how that performance might be achieved (i.e., via complexity or consciousness, etc.). Also this view allows one to compare various types of agents. For example, consider two agents (machines, robots, etc.) that have roughly the same **OP**, but the first needs much more *resources* (**R**) to do the task. Clearly, the second agent is preferable from a practical point of view.

Here, the following representative set of machine intelligent tests will be outlined [36, 38–50]:

- Text compression test
- Psychometric test
- C-test
- Universal intelligent test
- Symbolic machine intelligence test
- Multi-dimensional test

5.7.2 Text Compression Test

This is a binary pass or fail Turing-like test [36, 39]. The Turing test is simply replaced by a *text compression test*. This is similar to a test of a person's comprehension and knowledge in a domain by providing to him a passage text with missing words. In standard text compression, based on symbol frequencies, the compression obtained is relatively small. More efficient compressors are achieved by employing better models that capture higher-level linguistic features (e.g., grammar features). The best text compressors of this type for English can give compressed text to about 1.5 bits per character. A human who is able to use more general knowledge, the argument's logical structure, and so on, can achieve a better compression of about 1 bit per character. Thus it follows that compression estimation can provide a simple (and easily computed) measure of how good is a "language domain knowledge and inference model" of a machine, in comparison to a human. The similarity of this text compression test with Turing test can be seen using the "equivalence of compression and statistical prediction" [40]. Let a compression test which involves a large amount of dialogue. If a compressor is extremely good on such a test, then by the above equivalence it will be able to determine which sentences are likely to be at a certain point of the dialogue, and which are not. Therefore, since failure to pass a Turing test occurs when a machine (or human) generates a sentence which is importable for a human, extremely good performance on dialogue compression implies the ability to pass a Turing test. Text compression tests aim at measuring the ability of inductive inference (learning) with *minimum message length* (**MML**). MML advocates two part compression as a way to perform inductive inference and ultimately intelligence [41]. The problem of compression tests is the fact that it is unclear how to translate a powerful compressor into general machine intelligence.

5.7.3 Psychometric AI Tests

Psychometric tests of machine intelligence are based on the following view of intelligence: "An agent is intelligent if and only if its performance is superior at all established, validated tests of (human) intelligence, including tests of artistic and literary creativity, and mechanical skills [42]. These tests were used, for example, for building robots that can perform well on standard psychometric tests designed for humans (e.g., the Wechsler Adult Intelligent Scale). But, though these tests may work well with humans, they are not as efficient when applied to machines, since they are highly anthropomorphic. Humans are not mere collections of specialized algorithms designed only to answer specific IQ test questions like machines (robots). A robot might be able to obtain a respectable IQ score, even though outside the repertory of questions for which it was designed might be useless. Trying to face these limitations by expanding beyond standard tests is not a solution, since there remains the difficulty of what, and what not, to test for. In conclusion, psychometric AI (at least with its present development) does not address this problem satisfactorily.

5.7.4 C-Test

This test was devised in [43] with sequence-completion exercises similar to those found in some IQ tests. The **C-test** *(complexity test)* uses exercises created from computational principles and its complexity was mathematically quantified via constructs derived from Kolmogorov complexity theory (a theory similar to MML) [44]. The complexity c of the C-test is basically the number of the shortest program that generates the sequence. Clearly, for the C-test, the most difficult questions are the ones that are the most intelligent to solve. Therefore, it follows that we can construct a formal test of intelligence by formally defining and measuring the complexity of test problems. The idea of doing this was first coined by Chaitin [45]. The C-test was constructed by including exercises with complexity in the interval $7 < c < 14$ which was deemed suitable for humans. Complexity smaller than $c = 7$ was considered trivial, whereas $C > 14$ was very challenging for the experimenter. The test was applied to humans and the results were comparable to results of some IQ tests that were also administered to the same individuals. The C-test performed to humans indicated that intelligence is no longer only what the IQ tests measures, but (also) a convenient mathematical concept.

The C-test offers many advantages:

- It is formally motivated.
- It is precisely defined.
- It can perhaps be used to measure the performance of computers as well as biological systems (on the same scale) without any bias towards any particular structure or species.
- It is perhaps the only formal machine intelligence test that was successfully administered to humans.

5.7.5 Universal Intelligence Test

This test was devised using the *universal intelligence measure* (**UIM**) which is not anthropocentric, and embraces an extremely large gamma of capabilities from the simple reactive robotic agents to universally optimal agents, including terrestrial and even extraterrestrial life [43]. Here we will briefly review the

Universal Intelligence test of [36] which uses the concept of *reinforcement learning* that can be used for learning in unknown environments [46]. At each cycle of the agent-environment interaction the agent sends an action to the environment, which is then responding with an observation signal and a scalar feedback reward (positive or negative). The aim of the agent is to select its actions, on the basis of its past observations and rewards, such that to maximize the positive rewards received over time. Clearly, any problem can be formulated in this reinforcement learning setting (from playing a game, to adaptive robot control, and even writing an award winning novel).

Mathematically, the process of agent-environment interaction is formulated in terms of the probability π, which is a probability measure over actions conditioned on the current history, for example $\mu(a_3|o_1, r_1, a_1; o_2, r_2, a_2)$ where a_i is the action, o_i the observation, and r_i the reward at time step $i(|i = 0,1,2,\ldots)$.

The performance (intelligence) of agent π in environment μ, where μ is similarly defined, i.e., as $\mu(o_3, r_3 | o_1, r_1, a_1; o_2, r_2, a_2)$, is measured by its total expected reward (value):

$$V_\pi^\mu = E\left[\sum_{i=1}^{\infty} r_i \,|\mu, \pi\right]$$

To obtain a single performance measure V_π (not depending on a particular μ), V_π^μ is averaged over the largest class ε which is the class of all computable probability distributions (environments) μ, or equivalently, the set of all computable environments. Without any mathematical details the single performance measure, i.e. the universal *intelligence measure* (**UIM**), is given by [48]:

$$Y(\pi) = \sum_{\mu \in \varepsilon} 2^{-K(\mu)} V_\pi^\mu$$

where $K(\mu)$ is the length of the shortest program that computers μ (called *Kolmogorov complexity*).

By construction, UIM is a measure of the general ability of an agent to exhibit good performance in a very wide repertory of environments, and so it measures the intelligence in the general sense as proposed in [47]. Other important features of UIM are:

- It does not put any constraint on the internal workings of the agent. It suffices only to be an agent that can receive input (including reward signal) and produce output.

- It is a continuous performance measure, and so it gives more information than Turing and other tests for incremental progress.

The major drawback is that UMI is only a mathematical definition, which may not be suitable for evaluating real-word intelligent agents directly. The interested reader on an algorithmic (approximate) computation of UMI could refer to [48].

5.7.6 Symbolic Machine Intelligence Test

Symbolic machine intelligence tests (**SMI**) employ symbolic AI models which are based on the physical symbol system hypothesis. A symbolic AI model has the general form

$$X \underset{\rightarrow}{\mathbf{M}} Y$$

where X indicates the input elements, Y the output elements, and $\mathbf{M}$ a mapping that describes how X is mapped (transformed) to Y.

In the case of measuring the intelligence of machines, the set X represents the definitions of the variables, the set Y is the set of intelligent measures obtained, and $\mathbf{M}$ is the intelligent measurement methodology adopted. It is noted that the task of determining $\mathbf{M}$ is very difficult and depends largely of how the machine is defined. We look for a mapping (methodology) $\mathbf{M}$ which produces machine intelligent measures (MIQs), that have as input the set of intelligent machines, and as output one element of the set Y which possesses a *complete order* property (e.g., set inclusion property "C" or numerical order property "$\leq$", etc.).

Let A and B be arbitrary sets (crisp or fuzzy). We call a *relation* R a subset of the Cartesian product A $\times$ B. If A $=$ B, then this relation is said to be "*a relation in* A". In the following, a symbolic machine intelligence quotient *(symbolic MIQ)* will be derived for the case where the symbolic AI system belongs to the class of *Expert Systems* (**ES**) consisting of a set of knowledge base facts and rules [49, 50]. Consider an expert that has n knowledge base facts and rules of which m < n facts and/or rules do not offer new information (intelligence) to the system, and so they are redundant. In this case we have only n-m independent facts and rules, and the *symbolic MIQ* (**SMIQ**) can be defined as [37]:

$$\text{SMIQ} = 100 \times (n - m)/n$$

For example, consider an ES section which refers to a genealogical tree and has the following facts and rules:

r1. father (john, dennis)
r2. grandfather (john, jim)

r3. father (dennis, jim)
r4. grandfather (dennis, mark)
r5. father (jim, mark)
r6. IF father (X,Y) AND father (Y,Z)
    THEN grandfather (X,Z)

where father (X,Y) means that "the father of X is Y" and "grandfather (X,Z)" means that "the grandfather of X is Z". It is easy to see that elements r2 and r4 provide no additional information and they are redundant, since, e.g., the statements "dennis is father of john" and "jim is father of dennis" imply that "jim is grandfather of john". Thus the elements r2 and r4 can be eliminated since they are not independent of the other rules, and hence $\mathrm{SMIQ} = 100$ $(6 - 2)/6 = 66.26$. Note that the same type of SMIQ can also be used if our system is a neural network (connectionist system) for which a number of n neurons provide a good solution, but with another topology and type of neurons we can get the solution with only n-m neurons.

5.7.7 Multidimensional Machine Intelligence Test

This test is based on the ontological definition of machine intelligence which employs a number of machine features such as:

- Short-term learning
- Long-term learning
- User interface
- Recognition
- Optimization
- Scheduling/planning
- Reactivity
- Adaptation
- Robustness
- Autonomy

The multidimensional machine intelligence measure is then defined as a *weighted sum* of the individual MIQs of the machine's subtasks, and is calculated using a set of predefined weights that reflect the importance of each characteristic for the total machine intelligence [51]. Of course, a machine is not necessary to have all the above characteristics. The individual MIQs can be estimated by different methods. Let MIQ_i be the machine intelligence corresponding to the ith characteristic, then the total MIQ of the machine is given by:

$$MIQ = \left(\sum_{i=1}^{n} w_i MIQ_i \right) \left(\sum_{i=1}^{n} w_i \right)$$

where n is the number of characteristics considered, and w_i is the weight given to characteristic i.

A closely related measure to the multidimensional machine intelligence measure is a machine measure which refers to the ability of the machine to exhibit human-like performance with respect to a number of basic attributes such as:

- Kinetic intelligence
- Potential intelligence
- Functional convenience
- Physical friendliness

Each attribute can be represented by fuzzy variables which take linguistic values such as:

- Similar to human performance
- Very similar to human performance
- Lightly similar to human performance
- Difficult to use
- Easy to use

The degree of intelligence of the machine is also represented by fuzzy values for each attribute (e.g., enough intelligence, not enough intelligence), corresponding to fuzzy values of the MIQ (e.g., high, low MIQ). The technicalities of this fuzzy multi attribute MIQ are based on fuzzy reasoning and logic [52, 53]. The fuzzy values are obtained by assuming that MIQ is proportional to the machine's ability to mimic human behaviors, and reciprocal to the human operators intervention (effort) to operate the machine and achieve the desired performance [51].

5.8 Concluding Remarks

Over the years psychologists and other thinkers have made great efforts to understand, clarify and organize the complex mental phenomena of human intelligence. Analogous efforts were made by scientists and engineers to understand and build-in machines with intelligence. In this chapter we have studied fundamental philosophical aspects of both human intelligence and machine (computer/robot) intelligence. These aspects include human and machine intelligence ontology ("what is", "what are its forms", "how it is measured"), the historical landmarks of the study of machine intelligence, the Turing machine intelligence test, and a number of alternative machine intelligence tests (viz., text compression test, psychometric-like test, symbolic machine intelligence test, universal test, and multidimensional test).

Questions about intelligence that need more research are:

- Human intelligence is evolving, dynamic (always changing), always progressing, never ending, explosive and powerful. Can robot intelligence have these features? If yes, how can they be achieved?
- Can a robot have a free will? How can this be done? How much can this be done?
- How people perceive intelligent robots? How can the appearance, intelligence, autonomy, and conscious conscience modify humans' perception of intelligent robots?

References

[1] E. Hurt, Human Intelligence, New York, NY, U.S.A.: Cambridge University Press, 2010.

[2] N. Makintosh, IQ and Human Intelligence, Oxford/New York: Oxford University Press, 2011.

[3] R.J. Stenberg and S.B. Kaufman, Trends in Intelligence Research, Intelligence, 40, 2012, 235–236.

[4] R.J. Stenberg, Beyond IQ: A Triarchic Theory of Human Intelligence, New York, NY, U.S.A.: Cambridge University Press, 1985.

[5] A. Binnet and T. Simon, New Methods for the Diagnosis of Intellectual Level of Subnormality, In: E. Kite (Trans), The Development of Intelligence in Children, Vineland, NT, Publications of the Training School of Vineland, 1916 (Originally Published in L'Annèe Psychologique, 12, 1905, 191–244).

[6] C. Spearman, The Abilities of Man, New York, NY, U.S.A.: MacMillan, 1927.

[7] P.J. Hermstein and C. Murray, The Bell Curve: Intelligence and Class Structure in American Life, New York, NY, U.S.A.: Free Press, 1994.

[8] H. Gardner, The nine types of intelligence, Overview of Multiple Intelligences Theory, http://skyview.ransd.org/Ischmidt/Projects/TheNineTypesofIntelligence

[9] A. Turing, Computing Machinery and Intelligence, Mind, 59, 1950, 433–466.

[10] K. Krishna Kumar, Intelligent systems for aerospace engineering: An overview, NASA Technical Report (ID: 20030105746), 2003.

[11] S. Russel and P. Norvig, Artificial Intelligence: A Modern Approach, Upper Saddle River, NJ, U.S.A.: Prentice Hall, 1995.

[12] J.C. Bezdek, What is computational intelligence? In: J.M. Zurada, R.J. Marks II and C.J. Robinson (eds.), Computational Intelligence, Imitating Life, New York, NY, U.S.A.: IEEE Computer Society Press, 1994, 1–12.

[13] I.J. Rudas and J. Fodor, Intelligent Systems, International J. Computers, Communications and Control, III (Supp. Issue), 2008, 132–138.

[14] E. Rich, Artificial Intelligence, New York, NY, U.S.A.: McGraw-Hill, 1983.

[15] B.G.A. Buchanan, Brief History of Artificial Intelligence, AI Magazine, 26(4), 2005, 53–60.

[16] P. Schaumann, Short History of Computing: Artificial Intelligence and Robotics, http://philipps-welt.info/AI_history.htm

[17] A. Newell and H.A. Simon, Heuristic problem-solving: The next advance in operations research, Operations Research, 6(1), 1958, 1–10.

[18] A. Newell and H.A. Simon, Computer science as empirical inquiry: Symbols and search, Communications of the ACM, 19(3), 1976, 113–126.

[19] H. Dreyfus, What Computers Can't Do, Cambridge, MA, U.S.A.: MIT Press, 1972.

[20] H. Dreyfus, What Computers Still Can't Do, Cambridge, MA, U.S.A.: MIT Press, 1979.

[21] J. Lucas, Minds, machines and Gödel, In: A.R. Anderson, Minds and Machines, Upper Saddle River, NJ, U.S.A.: Prentice-Hall, 1964.

[22] D. Crevier, AI: The Tumultuous Search for Artificial Intelligence, New York, NY, U.S.A.: Basic Books, 1993.

[23] J. Searle, The Mystery of Consciousness, The New York Review of Books, New York, NY, U.S.A., 1997.

[24] H. Moravec, Bill Joy's Hi-Tech Warning, 2001. http://www.gigablast.com/get?q=&c=dmoz3dd=1180344

[25] R. Penrose, The Emperor's New Mind, Oxford U.K.: Oxford University Press, 1999.

[26] J.L. Noyes, Artificial Intelligence with Common Lisp, Lexington, MA, U.S.A.: Heath, 1992.

[27] R. Kurzeil, The Age of Spiritual Machines: When Computers Exceed Human Intelligence, Middlesex/New York, NY, 1999.

[28] G. Gilder and J. Richards, Are We Spiritual Machines? The Beginning of the Debate, In: J. Richards (ed.), Are We Spiritual Machine? Ray Kurzweil vs. the Critic of Strong AI, Discovery, Seattle, U.S.A.: Institute Press, 2002.

[29] W.S. Bainbridge, Progress toward Cyberimmortality, In: I. Basset (ed.), The Scientific Conquest of Death: Essays on Infinite Lifespans, Immortality Institute, Wausau, 2004.

[30] D.R. Hofstadter, Godel, Escher, Bach: A Eternal Golden Braid, New York, NY, U.S.A.: Basic Books, Inc., 1999.

[31] J. Hawkins and S. Blakeslee, On Intelligence, New York, NY, U.S.A.: Henry Holt and Company, 2004.

[32] J. Wang, On the limit of machine Intelligence, International J. of Intelligence Science, 3, 2013, 170–175.

[33] S. Harnad, Minds, machines and Searle, J. Theoretical and Experimental Artificial Intelligence, 1, 1989, 5–25.

[34] P. Schweizer, The truly total Turing test, Minds and Machines, 8, 1998, 263–272.

[35] S. Watt, Naïve psychology and the inverted Turing test, Psychology, 7(14), 1966.

[36] S. Legg and M. Hutter, Tests of Machine Intelligence, In: M. Lungarella, F. Iida and J. Bongard (eds.), 50 Years of Artificial Intelligence, Berlin: Springer, 2007, 232–242.

[37] J. Falqueto, W.C. Lima, P.S. Borges and J.M. Barreto, The measurement of artificial intelligence: An IQ for machines? Proceedings of IASTED International Conference on Modeling, Simulation, and Control, Innsbruck, Austria, Feb. 19–22, ACTA Press, 2001, 401–413.

[38] What is Intelligence? Machine Intelligence Research Institute (MIRI) http://intelligence.org/2013/06/19/what-is-intelligence-2/

[39] M.V. Mahoney, Text Compression as a test for artificial intelligence, Proceedings of AAAI/IAAI'99, American Association for Artificial Intelligence, 1999, 970.

[40] T.C. Bell, J.G. Cleary and I.H. Witten, Text Compression, Upper Saddle River, NJ, U.S.A.: Prentice Hall, 1990.

[41] C.S. Wallace, Statistical and Inductive Inference by Minimum Message Length, Berlin, Germany: Springer-Verlag, 2005.

[42] S. Bringsjord and B. Schimanski, What is Artificial Intelligence? Psychometric AI as an answer, Proceedings 18th International Joint Conference on Artificial Intelligence, 18, 2003, 887–893.

[43] J. Hèrnandez-Orallo, On the computational measurement of intelligence factors, Proceedings of Workshop on Performance Metrics for Intelligent Systems, Gaithersburg, MD, U.S.A., 2000, 1–8.

[44] C.S. Wallace and D.M. Bouldon, Minimum message length and Kolmogorov complexity, Computer Journal, 42, 1999, 270–283.

[45] G.J. Chaitin, Gödel's theorem of Information, International J. Theoretical Physics, 22, 1982, 941–954.

[46] R. Sutton and A. Barto, Reinforcement Learning: An Introduction, Cambridge, MA, U.S.A.: MIT Press, 1998.

[47] S. Legg and M. Hutter, A collection of definitions of intelligence, In: Advances in Artificial Intelligence, 157, Amsterdam, Netherlands: IOS Press, 2007.

[48] S. Legg and J. Veness, An approximation of the universal intelligence measure Proceedings of Ray Solomonov 85th Memorial Conference, LNAI, Melbourne, Australia, 2011.

[49] D. Waterman, A Guide to Expert Systems, Reading, MA, U.S.A.: Addison-Wesley, 1986.

[50] E. Post, Formal reductions of the general combinatorial problem, American J. of Mathematics, 65, 1943, 197–298.

[51] Z. Bien, W.-C. Bang, W. Yu and G.-Y. Lin, R&D trends on intelligent systems and MIQ as a performance measure. Available at: http://han dle.nef/10203/135725, EE Conference Papers; http://ctgate. kaist.ac.kr/bd_publications/pds/International_conf002.pdf

[52] L.H. Tsoukalas and R.E. Uhrig, Fuzzy and Neural Approaches in Engineering, New York, NY, U.S.A.: John Wiley, 1997.

[53] S.G. Tzafestas and A.N. Venetsanopoulos (eds.), Fuzzy Reasoning in Information, Decision, and Control Systems, Boston/Dordrecht: Kluwer, 1994.

6

Robot Autonomy

Control leads to compliance; autonomy
leads to engagement.
Daniel H. Pink

Self-determination could mean
independence, federal, and autonomy.
Jalal Talabani

At the end of the day, tech workers are not robots:
they feel, they think, and they have values.
Ed Lee

6.1 Introduction

Autonomy is a major philosophical and practical issue for human personal
and societal life and prosperity, and has several senses and interpretations in
present-day social, moral, and political discussions. The core idea of personal
autonomy is to have personal rule of the self, while being free from controlling
interference by others. An autonomous person behaves in accordance with a
freely self-chosen and informed plan, while a person of diminished autonomy
(e.g., a prisoner or other institutionalized person) is controlled (at least in some
respects) by others or is incapable of acting on the basis of his/her own desires
and plans.

In public health, autonomy refers to the autonomous decision making of a
patient, or subject of medical research, which is based on informed consent. All
current medical treatment and research codes of ethics include as a principal
component the informed-consent based autonomy. In government, autonomy
refers to one's own self-governance. In ethics an autonomous person acts

129

morally for the sake of doing "good" independently of other incentives. In general, rational autonomy refers to an agent's acting on the basis of what he believes to be the weighted reasons.

Actually, the concept of autonomy in robotics has a different interpretation than that of philosophy. In philosophy the interpretation of autonomy is focusing on the capacity to select goals for oneself. In robotics autonomy has a more technical interpretation which refers to the ability of independent execution of tasks for achieving goals. An issue which is of great importance in robotics is the understanding of what makes one's goals genuinely one's own. In robotics, autonomy may be at several hierarchical levels ranging from full autonomy (no human intervention) to non-autonomy (full human intervention), with intermediate levels of autonomy such as executive control, supervised control, shared control, assisted teleoperation, and action support.

The purpose of the present chapter is to study robot autonomy focusing on both ontological and epistemological philosophical issues. Specifically, the chapter discusses the following issues:

- What is autonomy (definition and types of autonomy).
- What is robot autonomy (definitions and features).
- Levels of robot autonomy.
- Philosophical considerations of robot autonomy (freedom of selecting goals, intrinsic ownership of goals, integration of body and control system, relation of autonomy with intelligence and capacity).

6.2 What Is Autonomy?

There are several conceptions of *autonomy*, all of which are based in one or the other way upon the core concepts of self-law, self-government, self-rule, or self-determination, but differ in the details [1–6]. The word autonomy comes from the Greek word "αυτονομία" (autonomia) which consists of two words: "αυτό" (auto = self) and "νόμος" (nomos = law/rule). It means self-governance, self-legislation, self-ownership, and personal sovereignty.

Historically, the word autonomy was predominantly used by ancient *Greek thinkers* to describe the right of city/states to self-legislate and govern free of external interference or force, although some philosophers of the time had also personal autonomy in mind when they wrote of persons being guided or ruled by reason (Plato, Aristotle). In *medieval* period the word autonomy was not in use although again many medieval philosophers were agreeing that those who live in accordance with right reason or the God's will are properly self-governed.

But it was not until the 19th century that autonomy came (in Europe) to refer also to the conduct of individuals, of course with different interpretations. The concept of autonomy as a philosophical issue was initiated by Kant, although it can be difficult to see how Kantian *"autonomy of the will"* is related to any sort of autonomy that is currently under debate. In other words, the question is: "what can the value of one's making one's own choices and leading his/her own life or his/her own right to do so, have to do with autonomy of the will as Kant interpreted it? In the 19th century a big contribution to the discussion of the normative significance of autonomy was made by John Stuart-Mill in his book *"On Liberty"*. His understanding of autonomy as self determination of mind continues up to the present times to have strong influence on discussions about normative significance of autonomy versus paternalism of several forms.

As argued in [2] there is, indeed, a deep connection between the "Kantian autonomy of the will" and autonomy in the sense of "one's making his/her own choices leading his/her own life, and his/her own right to do so". This connection is based on the idea that a right or claim to autonomy actually presupposes autonomy of the will. It assumes a capacity that persons have to impose demands rooted in the authority of free and rational will as such, and therefore in no value outside the will. In other words, claiming our autonomy commits us to assuming autonomy of the will too.

The principal modern conceptions of autonomy are [1, 5, 6]:

- Existentialist autonomy
- Personal autonomy
- Moral autonomy
- Autonomy as right or demand

A brief discussion of them is as follows.

Existentialist autonomy   This autonomy means the full freedom of persons to determine their natures and governing principles independently of any sorts of social, anthropological or moral designation. Thus possessing existentialist autonomy implies one's ability to choose one's nature without limitation from any principles not of one's own choice. Existentialist autonomy is a property that all human beings are said to have (sometimes not consciously), which may not be necessarily connected to morality or to rationality.

Personal autonomy   Personal or individual autonomy refers to one's autonomy in determining by his/her self most highly adopted values. It is a psychological ability, the possession of which enables individuals to reflect critically on their natures, preferences and goals, to locate the most authentic commitments, and to live in agreement with them in the presence of various kinds of internal and external inferences and influences. As such, personal autonomy is frequently set forth as an *ideal* for character or a virtue with its opposite being *blind conformance* (or "not being one's own person"). Personal autonomy and freedom (liberty) are typically used synonymously, but there are some important differences between them. Claiming that a person is *free*, is to claim that he/she is negatively free (i.e., not constrained by any internal or external forces that prevent making a choice and performing an action). But autonomy means the possession of a capacity to authentic living. Personal autonomy is actually the capacity to be the personal essence, one's fullness of *Being*. There is no fuzziness or uncertainty about autonomy when the personal essence is recognized as one's true being [7].

Moral autonomy   This is the autonomy of a person in making choices in accord with his/her own moral convictions or principles. It is the capacity of the will of a rational being to be a law to itself, independently of the influence of any property of objects or volition. Moral autonomy is associated with Kant who believed that autonomy of the will is both necessary and sufficient for the moral law.

Autonomy as right and demand   This refers to autonomy which is principally understood as a *right* or *demand* for self-determination (i.e., one has the right to determine his/her life without interference from social or political authorities or forms of paternalism). This form of autonomy establishes one's entitlement to several liberal rights (i.e., rights upon which our political status and freedoms are constructed) including the freedom to make decisions for oneself, free from interference. Of course the right to govern ourselves *(dejure autonomy)*, does not mean that we can actually do so in practice *(de facto autonomy)*. This means that possessing a formal right to self-government does not assure that it can be effectively exercised. Social status or economic power may either increase or diminish one's ability to realize his/her plans and achieve his/her goals. Generally, in all types of autonomy discussed above (personal autonomy, moral autonomy, autonomy as right) one can distinguish between a person's having the capacity to determine his conduct, and his/her successful exercise of this capacity. Autonomy as right or demand contrasts with the other ways of autonomy consideration, particularly as personal value or part of well-being or a benefit or a moral value in the Kantian sense. It may be regarded as falling within the frame of *preference-based versions of utilitarianism* [8]. This means that an important dimension of autonomy's value (in the broad sense) involves the right or claim or authority that persons have to demand their own choices and lead their own lives.

6.3 What Is Robot Autonomy?

Robot autonomy is distinguished in *"weak autonomy"* and *"strong autonomy"*. A robot is said to have *weak autonomy* if it operates without any connection to the outside. It is a purely *automated robot* which doesn't make choices for itself but involves embedded processing which repeats a fixed, pre-programmed sequence of commands (possibly very complex ones). In other words all possible desired courses of action are provided to the robot beforehand, and if the robot encounters unplanned situations it stops and call for human help. Therefore, for a weakly autonomous robot, choices may have been made and programmed already, or they have to be provided externally.

A robot is said to have *strong autonomy* (or *full autonomy*) if it can make choices on its own and is authorized to actually activate them. A fully autonomous robot tries to achieve its goals locally without human intervention, even when facing unanticipated events and internal or external uncertainties. An autonomous system that makes operational choices and operates in an intelligent human-like way, and can *"create"* solutions to ambiguous problems, is called an *"intelligent autonomous system"*.

As we have already seen, conceptually human autonomy is the capacity to be one's own person to live one's life according to reasons and motives taken as one's own and not the product of external forces. But autonomy in robots and machines should have a narrower nature compared to human autonomy, i.e., it should be considered *metaphorically*. Specifically, robot/machine autonomy cannot be defined absolutely, but only relatively to the goals and tasks or functions required. Of course, it may often happen that the results of the operations of a robot are not known in advance to human designers and operators. But this does not imply that the robot is a fully autonomous and independent agent in the human sense. Robot autonomy represents a class of *artificial autonomy*, which embraces all technological systems with embedded autonomy of various types and levels (operational autonomy, functional autonomy, full autonomy).

A representative list of definitions of robot (and agent) autonomy is the following:

Bekey "Autonomy refers to systems capable of operating in the real-world environment without any form of external control for extended periods of time" [9].

Wooldridge and Jennings "Autonomous agents operate without the direct intervention of humans or others, and have some kind of control over their actions and internal states" [10].

Murphy "*Functioning autonomously* indicates that the robot can operate, self-contained, under all reasonable conditions without requiring recourse to a human operator. *Autonomy* means that a robot can adapt to change in its environment (e.g., when the lights get turned off) or itself (a part breaks), and continue to reach a goal" [11].

Alami Autonomy means that "the robot should be able to carry out its actions and to refine or modify the task and its own behavior according to the current goal and execution context of its task" [12].

Antsaklis and Passino "Autonomy is the objective, and intelligent controllers are one way to achieve it...A highly-autonomous control system provides high adaptability to changes in the plant environment and the plan control objectives" [13].

Thrun "Autonomy refers to a robot's ability to accommodate variations in its environment. Different robots, possess different degrees of autonomy" [14].

Pattie Maes "Autonomous agents are computational systems that inhabit some complex dynamic environment, sense and act autonomously in this environment, and by doing so realize a set of goals or tasks for which they are designed".

Russel and Norvig "An agent is anything that can be viewed as perceiving its environment through sensors, and acting upon that environment through effectors" [15].

FIPA "An agent is a computational process that implements the autonomous communicating functionality of an application" (Foundation for Intelligent Physical Agents, www.fipa.org)

From the above definitions we can see that no single definition captures all features, capabilities, and possibilities of robot/agent autonomy. However, we can get a good view of autonomy by merging the features involved in these definitions.

These features and capabilities can be summarized as follows:

- No human (or other) intervention involved.
- Capability to refine or modify task, behaviors, or own operations.
- Capability of sensing and acting (perception and action).

- High-level adaptability and prediction ability.
- Capability to communicate functionality.
- Capability to work in uncertain, varying, and unpredictable environments.
- Capability to achieve goals. The philosophical issue of who poses the goals is not consistently considered in the literature (A discussion of this issue is provided in Section 6.5).
- Possession of several levels of autonomy.

In the following we discuss the issues concerning the "levels of robot autonomy".

6.4 The Levels of Robot Autonomy

The level of autonomy of a robot is specified by the extent of on-line supervision and intervention of the operator, programmer or designer in relation to the robot's operations in a varying environment. The influence of autonomy on human-robot and interaction-related issues is complex. Over the years much research was devoted for exploring this human-robot autonomy relation, particularly in critical areas such as unmanned systems and workload division in flight decks and other sophisticated automation systems.

A popular hierarchical taxonomy of robot autonomy suggested in the literature is the following, where in a top down way the level of autonomy is decreasing from full autonomy to weak autonomy and full manual teleoperation [16]:

- Full autonomy (no human intervention)
- Executive control
- Supervisory control
- Shared control
- Decision support
- Assisted teleoperation
- Action support
- Manual teleoperation (full human intervention)

The level of robot autonomy refers to all operations, namely *sensing*, *planning* and *action* (Figure 6.1), and specifies which aspects of the robot task have to be performed by the robot. This depends on the criticality and accountability of the task and the complexity of the environment. The variables/features involved are distinguished in:

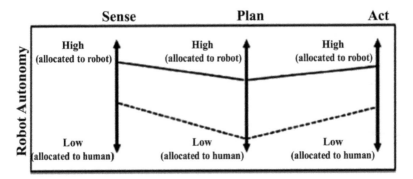

Figure 6.1   Levels of autonomy across the robot functions: sense, plan and act. Dotted line corresponds to assisted teleoperation, and solid line to semi-autonomous operation [16].

Source: http://scholarcommons.sc.edu/cgi/viewcontent.cgi?article=1127&context=csce_facpub

- *Robot related variables* (intelligence, learning, control, reliability, transparency).
- *Human-robot interaction variables* (social features, social effectiveness, speech interaction).
- *Human-related variables* (acceptance, situation, awareness, trust, work load).

Figure 6.1 shows how the levels of autonomy are distributed across the functions sense, plan, and act.

The problems of *task analysis* and *task allocation* are very important and their solution influences the actual overall properties of the designed system. For the task analysis problem there are available many popular techniques, but for the task allocation problem there are no generally accepted techniques. The selection of human-robot mix is typically done by empirical guidelines and qualitative criteria. A first solution attempt which was proved useful in practice is the so-called **Fitts** (or **MABA–MABA**) list of what *"men are better for"* and what *"machines are better for"*. This list is the following [17].

Men are better for:

- Detecting small quantities of visual and auditory energy.
- Perceiving patterns of light and sound.
- Improving and using flexible procedures.
- Storing information for extended periods of time and recalling appropriate parts.
- Carrying out inductive inference.
- Performing judgment.

Machines are better of:

- Responding quickly to control signals.
- Exercising large forces accurately and smoothly.
- Storing information briefly, and erasing it completely.
- Carrying out deductive reasoning.

This list was criticized by many researchers, since the human cannot be truly compared to the machine. Craik [18] has pointed out that we know how to replace a human with a machine only to the extent that human is understood as machine, while Jordan [19] says that "the main benefit of using the MABA–MABA list is that humans and machines are complementary".

A short description of the tasks performed by the *robot* (**R**) and the *human* (**H**) in the robot autonomy taxonomy, proposed in [16], is as follows.

Full autonomy (R, R, R) The robot/machine does everything (*sense :* R, *plan :* R, *act :* R) by ignoring the human.

Executive control (R, (H)/R, R) The robot autonomously senses the environment, sets plans, and implements action. The human gives abstract high-level goals (such as go to a specific location) [22, 23].

Supervisory control (H/R, R, R) The robot carries out all aspects of the task under continuous monitoring by the human, who can intervene and set a new goal or plan [24].

Shared control (H/R, H/R, R) The robot performs all aspects of the task (sense, plan, act) and asks for human help if it encounters difficulties, or the human may intervene and influence the robot with new goals/plans if necessary [25, 26].

Decision support (H/R, H/R, R) Both the human and the robot sense the environment and generate a plan, but the human chooses the task plan and commands the robot to execute the action [27].

Assisted teleoperation (H/R, H, H/R) The human assists the robot in all aspects of the task, but the robot senses (locally) the environment and can intervene (e.g., in order to avoid a collision) [28].

Manual teleoperation (H, H, H) The human does everything, namely sensing the environment and monitoring the system, generating plans, alternatives and goals, and implementing/executing the required actions [29].

Sheridan [17, 20] has proposed a set of guidelines for human computer task allocation between the two extremes using a working scale of computerization/autonomy from zero to one hundred percent, that helps very much the designer to do this selection (see also [21]). This working scale is shown in Table 6.1.

Table 6.1   Sheridan's scales of degrees of automation

1. The computer does nothing. Everything is performed by the human.	
2.	i. The computer gives the full set of alternatives of action, or
	ii. Reduces the alternatives to a few, or
	iii. Proposes a single alternative, and
	iv. Executes this alternative, or
	v. Gives the human a certain period of time to veto, before switching to automatic operation, or
	vi. Executes automatically, then necessarily informs the human, or
	vii. Informs the human only if asked, or
	viii. Informs the human if it decides by its own
3. The computer does everything and acts fully autonomously, by ignoring the human	

A summary of Sheridan's guidelines is as follows:

- *Try to find and remove any unnecessary constraints* which make the task allocation more difficult (going back to task analysis).
- *Try to find obvious allocations* (e.g., tasks that are easy to computerize must be allocated to the computer, most non repetitive tasks must be allocated to human, etc.).
- *Try to examine the extremes* A pro-computer extremist allocates to the human only the tasks that cannot be done by the computer. Conversely, a pro-human extremist supports the human with computer help only for the tasks that there is no other choice.
- *Use a proper human-computer task allocation between the two extremes (Table 6.1)*
- *Examine how fine an allocation makes* sense. It must be taken into account that task allocation to human to an excessively fine degree makes no sense, because the human cognition and attention cannot be portioned arbitrarily, or just turned on or off. Human memory tends to be less good in details, whereas it is better for more complete pictures and patterns.
- *Select between trading and sharing* Trading occurs when the human and computer perform their tasks serially, each handling the task back to the other when finished with only one part. Sharing occurs when the human and the machine work on the task in parallel (at the same time). Human and machine can cooperate in both modes. The designer must select the best on the basis of the task context.
- *Use many salient criteria* To this end, the designer must write down the available criteria for judging when one allocation is better than another, and try to rank order them in a suitable way, possibly using convenient quality weights.

A popular criterion, that helps the designer to decide when to automate and use supervisory control, is the *time needed to perform a task* by the human and the computer. In general, the time to execute a task increases as the complexity of the task increases. The time required to perform a task by supervisory control is the sum of the time spent for planning and teaching the task to be done, and the time to execute the task. Supervisory control is faster than manual control, only when the complexity of the task is such that the speed of execution by the computer (the machine) overbalances the planning and teaching overhead.

The *National Institute of Standards and Technology* (**NIST**, U.S.A.) has developed a *Framework for "Autonomy Levels for Unmanned Systems* (**ALFUS** Framework)" [30–32]. The ALFUS general view, and the autonomy levels mapping across the HRI-levels continuum are shown in Figures 6.2 and 6.3.

The ALFUS framework report involves the following issues:

- *Terms and definitions* (a set of standard terms and definitions).
- *Detailed model for autonomy levels* (a set of three comprehensive and detailed metrics, together with a set of processes for determining the autonomy).
- *Summary model for autonomy levels* (a concise scale and a set of corresponding definitions for the autonomy levels).
- *Guidelines, processes, and use cases* (a white paper for describing the conversion of the detailed technical ALFUS model into the summary model, together with guidelines for the application of the generic framework to specific ALFUS applications).

The variety of autonomous systems that exist, and are expected to be developed, makes a common set of terminology and definitions with proper metrics of paramount importance for all unmanned systems' domains (aerial: **UAV**, ground: **UGV**, underwater: **UUV**, and surface: **USV**, etc.).

The ALFUS model characterizes the autonomy levels through the following three perspectives (or axes of autonomy scores):

- *Mission complexity* (subtasks, decisions, commanding structure, interfaces, planning, allowed latencies, situation awareness).
- *Human interface* (frequency, duration and robot initiation; workload and skills, operator to unmanned system ratio %).
- *Environmental difficulty* (static: terrain, soil; dynamic: object frequency, density, type; EMI; urban, rural, weather; operational: threats, decoy, mapping).

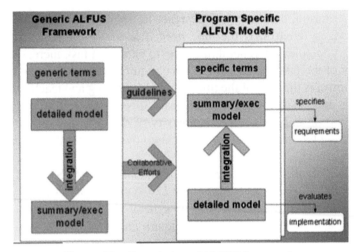

Figure 6.2   General view of ALFUS framework.

Source: http://www.nist.gov/el/isd/ks/upload/ALFUS-BG.pdf

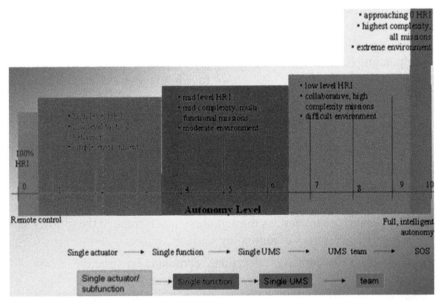

Figure 6.3   ALFUS autonomy levels and their mapping to human-robot interfaces (HRI), i.e., high-level HRI, mid-level HRI, and low-level HRI.

Source: http://www.nist.gov/el/isd/ks/upload/ALFUS-BG.pdf

The possible options involve weighted average and weighted maximum/minimum options. The ALFUS framework has been represented in an appropriate 3-dimensional pictorial form. A tool has been developed for automatically computing the autonomy levels based on the weights and metric scores entered by the user. The subtask scores of each level are evaluated (starting from the lowest level tasks) against all the three sets of the ALFUS metrics, and are then weighted and averaged to give a composite score for the subtasks. The subtask scores are further weighted and averaged to provide the composite scores for the next higher level tasks. This process is repeated until the overall mission obtains its composite score.

A discussion of the challenges and opportunities for putting autonomous systems on-board spacecraft is provided in [33]. These include maintenance of the spacecraft despite failures or damage, enlargement of the scientific group via *"virtual presence"*, and cost-effective operation over long periods of time.

A road map of autonomous service robotics was provided by the SRI Consulting Business Intelligence (Disruptive Technologies Global Trends 2005). This road map that extends from the 2000's and goes toward 2030, and more, is shown in Figure 6.4.

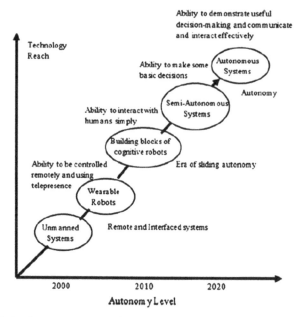

Figure 6.4 Technology-autonomy roadmap of autonomous service robots, including medical, assistive, and social robots (SRI Consulting Business Intelligence).

This map shows the timing, features, and applications of significant technology milestones that would be necessary to produce efficient autonomous systems for service and socialized robots. Automation (autonomy issues regarding air traffic/transportation systems are discussed in [34, 35].

6.5 Some Philosophical Considerations of Robot Autonomy

In Sections 6.3 and 6.4 we have discussed the concept of robot autonomy from a rather technical point of view presenting a number of existing definitions and the levels of autonomy. In this section we will discuss a little more robot autonomy from a philosophical point of view.

The issues to be discussed are:

- Freedom of selecting goals.
- Intrinsic ownership of goals.
- Integration of body and control system.
- Relationship of autonomy with intelligence and capacity.

6.5.1 Freedom of Selecting Goals

Robots can operate *independently* (even *"freely"* choosing how to act in order to achieve goals) but the goals they are trying to achieve are set (at the moment) by humans. Although on-line human intervention has been considerably reduced, there is still enormous amount of human off-line involvement before the robot's functioning.

Philosophically, *freedom of will* of humans is naturally *strong* (e.g., I'd like to eat sweets but since I want to loose weight I eat fruits instead), but humans can also have *weakness of will* (e.g., I want to be healthy and think that drinking beer is harmful for my health but I get beer in a shop and drink it all the same), or *brainwashed will* (e.g., external influences make me think that I want to drink beer because it is good for me, so I decide to sneak a beer-factory at night and drink the whole lot). From the above it is clear that in humans we have [36]:

- *Autonomy of will* (selecting a goal) and *action* (performing a specific action).
- *Autonomy of action* (weakness of will).
- *No autonomy at all.*

Clearly, the responsibility for our actions decreases as we go from (strong) autonomy, to autonomy of action, and non autonomy. From a philosophical point of view robots may be considered to be in a situation similar to that of humans who are brainwashed, and so robots cannot be regarded as having any serious degree of autonomy. In other words, philosophically robots are not autonomous, because they don't choose their own goals, and they are not even aware that their goals are set by humans. So, we can speak about robot autonomy only metaphorically. But it should be noticed that the diachronic debate about the issues of autonomy and the ability to choose one's goals is still ongoing leading to different interpretations of autonomy. Although philosophy has started with a conception of agent causation and the capacity to freely choose goals, it ended up in the 18th century with a strong mechanistic view. To see this evolution in the issue of autonomy and free will we mention just a few landmarks.

Descartes (1641) has equated the will with freedom of choices. He said: "The will simply consists in our ability to do or not do something" [37].

Hume (1739) argued that "the acts of will are caused by motives according to the bonds of necessity" [38].

Hartley (1749) claimed that there is a natural mechanism that drives our actions, saying: "... each action results from the previous circumstances of body and mind ..., so that a person cannot do indifferently either of the action 'A', and its contrary 'a', while the previous circumstances are the same, but is under an absolute necessity of doing one of them, and that only" [39].

Skinner (1971) has expressed the most extreme view that the ultimate causes of behavior lie outside the agent concluding that *"man is not free"* [40].

Actually, the issue of "freedom of will" is one of the major debate issues of philosophy, and so, naturally, one should not expect that research in robotics could solve a metaphysical problem which philosophy itself is still trying to resolve. The technical problem of creating robots with greater independence of action (automation) does not deal with the philosophically important aspects of autonomy. These two different views of autonomy (independence and autonomy) do not provide a solid ground for a debate between robotics and philosophy [36]. Thus, from this perspective robots should not be expected to add new knowledge about autonomy.

Nevertheless, robotics may profit from a philosophical analysis of the concept of autonomy which may help to get a more complete specification of the conditions needed to be fulfilled by robots before they are characterized as "autonomous". Similarly, philosophy might gain in its effort to clarify what is (and is not) meant by autonomy, by exploiting several challenging examples of robotic behavior as test cases.

6.5.2 Intrinsic Ownership of Robot Goals

The issue of intrinsic ownership of goals has to do with the question: "how and when goals of agents genuine become theirs? This issue must be distinguished from the harder issue of "freedom of will" (i.e., whether an agent is free to choose what it wants to do, and how this would be compatible with physical determinism). In other words, when and on what basis, we could argue that robots are pursuing goals on their own? What makes my goals mine, even if I don't select them freely? Often, I decide to act towards achieving certain goals not because others impose them upon me or require me to achieve, but because they matter to me. In this case the goals become mine. In general, goals belong to an agent (robot, organism) when they arise out of the ongoing attempt, sustained by both the body and the control system to regulate its internal variables within the bounds of stability for the agent (i.e., when they are sustained by the so-called *"homeostasis"*). Therefore, integration of robot and its control system such that to guarantee "homeostasis", is fundamental for the intrinsic ownership of its goals [36].

6.5.3 Integration of Body and Control System

According to the biologist *Uexküll* (1957) the integration of all the components of an organism into one purposeful whole is an important aspect for their life. He said: "We no longer regard animals as mere machines, but as subjects whose essential activity consists of perceiving and acting", and emphasized the fact that machines act according to plans of their human creators, while living organisms are acting plans [41]. *Chiel* and *Beer* (1997) have stressed that the brain and the body have developed in continuous conjunction over the evolution and interaction with the environment [42]. Evolutionary robotics which aims to drive the robot designer and programmer *"out of the robot"* as much as possible is an interesting development in "automatic design via artificial evolution". Artificial evolution is based on computational *genetic algorithms* (**GA**). The genotypes of the robots of concern are subject to crossover with other genotypes, and further mutation. The candidates for reproduction are selected using a proper fitness criterion which measures the success of the robots on a specific task. In principle, the role of the robot designer may be limited to the specification of the fitness (selection) criterion, but in practice the role of the designer is much more than that. Many other issues of the genetic-algorithm/evolutionary process are selected by him (e.g., the number of individuals in the population, the rates of mutation and crossover, the length of the lifetime of each individual, etc.). On the other hand, very frequently the experimenter needs to interfere directly in order to resolve local minima and bootstrap problem issues [43].

It is clear that evolutionary robotics is relevant to robot autonomy, because there is less need for the designer or programmer's direct involvement in evolving artificial creatures like robots. But much research is still required before the achievement of practical autonomous robots via robotic evolution. This field is still in its infancy and the types of robots already created are fairly simple [44]. No doubt, "building the operator/controller into the body of the robot" is a deep and legitimate biologically inspired challenge to robotics towards creating autonomous robots, with *"ownership of goals"*, which deserves intensive research.

6.5.4 Relation of Autonomy with Intelligence and Capacity

In the literature, the terms intelligence, autonomy and capacity are often used in a confused way. In the discussion that follows we attempt to provide some light towards clarifying the relation of these concepts along the lines of [45].

Given that modern robots work in dynamic and uncertain environments, we can say that intelligence is the robot ability to determine the behavior which will maximize the *likelihood* of goal satisfaction, i.e., intelligence is *goal oriented* [46]. *Capacity* is the ability to successfully execute behaviors or actions in dynamic and uncertain environments. It is defined without any reference to the correctness or suitability of actions, and is concerned only with the successful execution of the required actions. On the contrary, intelligence is defined by the determination of correct behaviors, and has nothing to do with the execution of behaviors. Performing successfully a wrong action does not imply reduced capacity.

The above observations make clear that, actually, *intelligence* and *capability* are *independent*, i.e., one can have high-level intelligence and lack capability, and vice versa. In other words, intelligence and capability define an orthogonal design space (with axes the intelligence and capability) in which robot autonomy should be considered. In the transition of intelligence from human intelligence to robot intelligence, there is given more emphasis on the functional aspects, rather than on structural ones. Intelligence is typically measured via factor-based performance metrics, i.e., via how well an intelligent system can develop solutions to problems in changing and uncertain environments. A common mathematical measure for this is the likelihood (probability) of goal satisfaction.

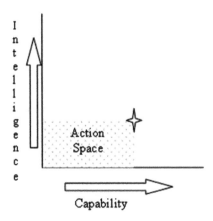

Figure 6.5   Autonomy range in the intelligence-capability space [45].

Autonomy is independent of intelligence, because autonomy is not addressing the quality of its decisions. An agent can be autonomous and stupid, or autonomous and intelligent. A system can make good decisions and execute them poorly or well.

Given that autonomy is the ability of an agent to decide its own behavior and execute that behavior, how does that relate to an intelligent agent such as a robot? A critical issue is that the agent must be capable to select between several options. If an agent has no choice in the behaviors that exhibits, then it cannot be autonomous. A prerequisite for exhibiting "autonomy" is the existence of behavior or action options. If an agent has options to select, and the ability to select and implement one of them, then it is autonomous [47]. Clearly, the inability to produce a behavior (course of action) precludes its use, and the inability to execute a course of action prevents it from being valid. Therefore, intelligence and capability provide upper bounds on the autonomy of an agent, i.e., intelligence and capability seem to define an action space in which autonomy is bounded. As intelligence and capability increase, the repertory of available options increases. The agent (robot) may be autonomous in which case it has the ability to select from these options, or non-autonomous in which case its action collapses to a single point for any goal-environment situation (Figure 6.5 [45]).

Here the following questions arise:

- What if an agent has several options, but always selects the same option?
- Is this different from an agent that has available only this particular option?

- Does this mean that one agent is autonomous and the other not?
- How can an external observer discriminate one agent from the other?

Intelligence and capability are independent in the abstract sense. In theory, no constraints exist that limit either range. An agent can be exceptionally intelligent but totally incapable to achieve any goal, or totally stupid and have exceptionally advanced execution capabilities [45]. In nature (ecology, biology) there is always a balanced combination of intelligence and capability. A successful balance is required for the survival of species and living organisms. In a highly competitive environment any increase in intelligence which is not matched by a proper increase of capability could lead to the extinction of a species. For example, consider two species in the same ecosystem which are confronted with a new food source. One of the species constructs an intelligent feeding strategy that can implement. The other species develops a feeding strategy that it is not capable to execute. They have both external energy coming up with an intelligent solution, but only the first has received a reward. In an evolutionary environment the species that come up with an executable strategy has a major competitive advantage, which, if it is sufficiently high, may lead to the extinction of the competitive species. In animals, intelligence and the capacity to exploit this intelligence must develop jointly; any imbalance between intelligence and capacity would lead to competitive failure of the species. Such a balance is also required in robots. A general rule in robot design is to limit intelligence to that which is sufficient for solving a given problem, and add more capability for implementing the solution. This will result in an improvement of success. Of course adding more intelligence will lead to greater improvement.

6.6 Concluding Remarks

In philosophy, human autonomy is interpreted as autonomy focusing on the capability to select goals for one self. In robotics, autonomy has a more technical interpretation which refers to the capability of independent execution of tasks for achieving goals. Robot autonomy may be at several hierarchical levels ranging from full autonomy (no human intervention), to full non-autonomy (full human intervention), with several intermediate levels of autonomy.

In this chapter we have studied robot autonomy, covering both ontological and epistemological philosophical issues. In particular, the following aspects were examined: (i) what is human autonomy, (ii) what is robot autonomy (definitions and features), (iii) levels of robot autonomy, and (iv) other philosophical considerations of robot autonomy (freedom in goal selection, integration of body and control system, and relation of autonomy with intelligence and capacity).

Questions about robot autonomy that require further research include:

- How human trust changes over the levels of autonomy, and how robot learning new tasks over time helps in the increase of human trust to the robot?
- How the type of robot's feedback about its state and autonomy (communication, gestures, emotion, etc.), across the levels of autonomy help in an effective and beneficial human-robot interaction and symbiosis?
- What type of safety systems are needed for personal and mental robots that function in contact or close approximation to humans?
- What is the impact of a mismatch between perceived and actual autonomy to human performance?

References

[1] B. Berofsky, Liberation from Self: A Theory of Personal Autonomy, Cambridge, U.K.: Cambridge University Press, 1995.

[2] S. Darwall, The Value of Autonomy and Autonomy of the will, Ethics, 116(2), 2006, 263–284.

[3] T. Hill, The Kantian conception of autonomy In: J. Christman (ed.), The Inner Citadel: Essays on Individual Autonomy, Oxford, U.K.: Oxford University Press, 2003.

[4] R. Young, The value of autonomy, The Philosophical Quarterly, 32(126), 1982, 35–44.

[5] V. Ashley, Philosophical models of autonomy, Essex Autonomy Project: Green Paper Report. http://autonomy.essex.ac.uk/philosophical-models-of-autonomy

[6] M. Piper, Autonomy: Normative, Internet Encyclopedia of Autonomy. http://www.iep.utm.edu/aut-norm/

[7] A.H. Almaas, Autonomy-Glossary. http://www.ahalmaas.com/glossary/autonomy

[8] J. Stuart Mill, Utilitarianism, London, U.K.: Longmans Green & Co,. 1879.

[9] G.A. Bekey, Autonomous Robots: From Biological Inspiration to Implementation and Control, Cambridge, MA: MIT Press, 2005.

[10] M. Wooldridge and N.R. Jennings, Intelligent Agents: Theory and Practice, Knowledge Engineering Review, 10, 1995, 115–152.

[11] R.R. Murphy, Introduction to AI Robotics, Cambridge, MA, U.S.A.: MIT Press, 2000.

[12] R. Alami, R. Chatila, S. Fleury, M. Ghallab and F. Ingrand, An architecture for autonomy, International J. Robotics Research, 17(4), 1998, 315–337.

[13] P.J. Antsaklis and K.M. Passino (eds.), An Introduction to Intelligent and Autonomous Control, Boston/Dordrecht: Kluwer, 1993.

[14] S. Thrun, Toward a framework for human-robot interaction, Human-Computer Interaction, 19(1–2), 2004, 9–24.

[15] S.J. Russel and P. Norvig, Artificial Intelligence: A Modern Approach, Upper Saddle River, N.J.: Pearson Education, Inc., 2003.

[16] J.M. Beer, A.D. Fisk, and W.A. Rogers, Toward a Psychological Framework for Levels of Robot Autonomy in Human-Robot Interaction, Technical Report HFA-TR-1204, School of Psychology, Georgia Institute of Technology. https://smartech.gatech.edu Also: Journal of Human-Robot Interaction, 3(2); 2014.

[17] T.B. Sheridan, Task analysis, task allocation, and supervisory control, In: M. Helander, T.K. Landauer, and P. Prabhu (eds.), Handbook of Human-Computer Interaction, Amsterdam, The Netherlands: North-Holland, 1997, 87–105.

[18] K.J.W. Craik, Theory of the human operator in control systems: Part 1, The operator as an engineering system, British J. Psychology, 38, 56–61: Part 2, Man as an element in a control system, ibid, 38, 142–148, 1947.

[19] N. Jordan, Allocation of functions between man and machines in automated systems, J, Applied Psychology, 47, 1963, 161–165.

[20] T.B. Sheridan, Supervisory control, In: G. Salvendy (ed.), Handbook of Human Factors and Ergonomics, New York, U.S.A.: Wiley, 1997, 1295–1237.

[21] S.G. Tzafestas, Human and Nature Minding Automation: An Overview of Concepts Methods, Tools and Applications, Dordrecht, The Netherlands: Springer, 2010.

[22] D.J. Bruemmer, D.A. Few, R.L. Boring, J.L. Marble, M.C. Walton, and C.W. Nielsen, Shared understanding for collaborative control, Proceedings of IEEE Conf. on Systems, Man, and Cybernetics, 35(4), 2005, 494–504.

[23] M. Baker and H.A. Yanco, Autonomy mode suggestions for improving human-robot interaction, Proceedings of IEEE Conf. on Systems, Man, and Cybernetics, 3, 2004, 2948–2953.

[24] D.B. Kaber, E. Onal, and M.R. Endsley, Design of automation for telerobots and the effect on performance operator situation awareness and

subjective workload, Human Factors and Ergonomics in Manufacturing, 10(4), 2000, 409–430.

[25] B. Sellner, F.W. Heger, L.M. Hiatt, R. Simmons, and S. Singh, Coordinated multiagent teams and sliding autonomy for large-scale assembly, Proceedings of the IEE: Special Issue on Multi-Robot Systems, 94(7), 2006, 1425–1444.

[26] T.-J. Tarn, N. Zi, C. Guo, and A.K. Bejczy, Function-based control sharing for robotics systems, Proceedings of IEEE/RSJ Intern. Conference on Intelligent Robots and Systems (IROS'95), 3, Munich, Germany, 1995, 1–6.

[27] D.B. Kaber and M.R. Endsley, The effects of level of automation and adaptive automation on human performance, situation awareness, and workload in a dynamic control task, Theoretical Issues in Ergonomics Science, 5(2), 2004, 113–153.

[28] L. Takayama, E. Marder-Eppstein, H. Harris, and J.M. Beer, Assisted driving of a mobile remote presence system: System design and controller user evaluation, Proceedings of IEEE Intern. Conference on Robotics and Automation (ICRA'2011), Shanghai, China, 2011, 1883–1889.

[29] P. Milgram, A. Rastogi, and J.J. Grodski, Telerobotic control using augmented reality, Proceedings of IEEE Intern. Workshop on Robot and Human Communication (ROMAN-95), Tokyo, Japan, 5–7 July, 1995, 21–29.

[30] H.-M. Huang, Autonomy levels for unmanned systems (ALFUS) framework, Proceedings of the National Institute of Standards and Technology (NISTSP), Gaithsburg, U.S.A., 2004.

[31] H.-M. Huang, K. Pavek, J. Albus, and E. Messina, Autonomy levels for unmanned systems (ALFUS) framework: An update, Proceedings of 2005 SPIE Defense and Security Symposium, Orlando, FL, U.S.A., 2005.

[32] H.-M. Huang, K. Pavek, B. Novak, J. Albus, and E. Messina, A framework for autonomy levels for unmanned systems, Proceedings of AUVSI's Unmanned Systems North America 2005, Baltimore, MD, U.S.A., 2005, 1–8.

[33] C.R. Frost, Challenges and opportunities for autonomous systems in space, Proceedings of National Academy of Engineering: U.S. Frontiers of Engineering Symposium, Armonk, New York, Sept. 23–24, 2010.

[34] G.L. Donohue (ed.), Air Transportation Systems Engineering, American Inst. of Aeronautics and Astronautics, Reston, 2001.

[35] M.S. Nolan, Fundamentals of Traffic Control, Pacific Grove: Books Cole Publ. Co., 1999.

[36] W.F.G. Haselager, Robotics, philosophy, and the problems of autonomy, In: I. Dorf (ed.), Cognitive Technologies and the Pragmatics of Cognition, London, U.K.: John Benjamins, 2007, 61–77.

[37] R. Descartes (1641), Meditations on first philosophy, In: J. Cottingham, R. Stoothoff, and D. Murdoch (eds.), The Philosophical Writings of Descartes (Vol. 2), Cambridge: Cambridge University Press, 1984, 3–62.

[38] Dottume (1739), A Treatise of Human Nature, Oxford, U.K.: Clarendon Press, 1978.

[39] D. Hartley (1749), Observations on Man, In: R. Brown (ed.), Between Hume and Mill: An Anthology of British Philosophy 1749–1843, New York, NY, U.S.A.: The Modern Library, 1970, 5–92.

[40] B. Skinner, Beyond Freedom and Dignity, New York, NY, U.S.A.: Bantham Books, 1971.

[41] J. von Uexküll, A stroll through the worlds of animals and men: A picture book of invisible worlds, Semiotica, 89(4), 1957.

[42] H.J. Chiel and R.D. Beer, The brain has a body: Adaptive behavior emerges from interaction of nervous system, body, and environment, Trends in Neurosciences, 20(12), 1997, 553–557.

[43] M. Mitchell, An Introduction to Genetic Algorithms, Cambridge, MA, U.S.A.: MIT Press, 1996.

[44] J.B. Pollack, H. Lipson, G.S. Hornby, and P. Funes, Three generations of automatically designed robots, Artificial Life, 7(3), 2001, 215–223.

[45] J.P. Gunderson and L.F. Gunderson, Intelligence $\neq$ autonomy $\neq$ capability, Proceedings of Performance Metrics for Intelligent Systems, Gaithersburg, MD, U.S.A., August, 2004.

[46] J.S. Albus and A. Meystel, Engineering of Mind, New York, NY, U.S.A.: Wiley Interscience, 2001.

[47] K.S. Barber and C.E. Martin, Agent autonomy: Specification, measurement, and dynamic adjustment, Proceedings of the Autonomy Control Software Workshop: Autonomous Agents, 99, Seattle, U.S.A., May, 1999, 8–15.

7

Robot Consciousness

Γνῶθισ᾽αυτόν *(Gnothi sauton)*
Know Thyself
Σωκράτης – **Socrates**
(Inscribed on the entrance wall of
Apollo's Temple at Delphi, Greece)

No computer has ever been designed that
is ever aware of what it is doing; but most
of the time, we aren't either.
Marvin Minsky

No man teach another self-knowledge.
He can only lead him or her up to self-discovery
—the source of truth.
Burry Long

7.1 Introduction

Consciousness fascinated humans from immemorial times. Scientists have attempted to find where is the source of consciousness, starting from the brain and going down up to the genetic code, but even at the genetic code level the trace of consciousness is not found, although all of us experience its existence. Consciousness is a more complex phenomenon than intelligence, and does not seem to be purely mechanical or chemical as materialists argue. If consciousness had been composed of matter only, it could be emulated by computing or other technological means. Scientists have made tremendous efforts to do this, but till now they were not able to generate complete machine (artificial) intelligence and consciousness. What they have done is to simulate the body by hardware in one form or another, and tried to simulate the mind and intelligence providing better and better approximations which however still lag far behind the desired ideal target.

Three principal philosophical problems about consciousness are the *existence problem* (why consciousness exists?), the *intelligibility problem* (how do we explain and understand consciousness?), and the *problem of explaining brain-mind correlations* (what possible explanation could there be for the way brain processes and sensations are correlated?). As argued by Nicholas Maxwell[1] the first problem has no solution. The second problem can be solved if we recognize that there is an explanation as to why science cannot explain consciousness, and that consciousness can be explained by *"empathic"* or *"personalistic"* explanation, compatible, but not reducible, to scientific explanation. The third problem can be solved by using Chalmer's *"principle of structural coherence"*. It involves assuming that our sensations (auditory, visual, tactile, etc.) correspond to minute scattered regions of the multidimensional space of all possible sensations that vary linearly and smoothly over that space. There is also the space of all possible scientist brain processes, and a unique one-to-one mapping (i.e. a common structure or mathematical isomorphism) between the space of sensations and the space of brain processes. This matching of structure is supposed to be *unique*, and if this assumption is valid, we have an explanation as to why brain processes and sensations are correlated as they are. They have to be correlated in this way because if they are not, the matching of structure, of unity based on continuity between sensations and brain processes would be lost. If this is true, then the third philosophical problem of consciousness has been solved. Nicholas Maxwell has put forward some possible objections to this postulate, which suggest that more theoretical and practical research work is needed to validate the unique-matching theory of brain and sensing spaces.

The purpose of this chapter is to provide a tour to the basic philosophical and practical issues of machine consciousness. Specifically the chapter:

- Discusses the ontological issues of consciousness through the views of several well known workers in the field.
- Investigates the same issues of machine-robot consciousness (purpose, plausibility, features).
- Presents several sets of requirements for conscious machines (Baars, McCarthy, Aleksander, Block, Lycan, Chrisley).

---

[1] Open J. Philosophy, 1(1), 2011, 1–10.

- Outlines a number of important practical computational models of machine consciousness (IBCA, CLARION, ACT-R, CERA, MANN).
- Discusses the issue of machine consciousness testing and measurement, and outlines the well-established *Con Scale* (Consciousness Scale) taxonomy which can serve as an engineering roadmap for the creation of conscious machines.

7.2 What Is Consciousness?

Consciousness has been over the years the subject of psychology, philosophy, cognitive science, neuroscience and other natural and social sciences. It involves everything we experience from a natural landscape, colors and noises to darkness and silence. Like intelligence, consciousness has been defined in several ways from a philosophical, psychological and scientific perspective. For their consciousness (as well as their conscience) human beings have possibilities incomparably better than those of animals even including the most clever ones. The term consciousness has its root to the Latin word *"conscious"* (con = with/together; scio = to know) which means *"knowing with"*. The Latin *"conscious libi"* means knowing with oneself, and the word *conscientia* means moral conscience.

The *Oxford English Dictionary* defines consciousness in a compact philosophical way as: "The state or faculty of being conscious, as a condition and concomitant of all thought, feeling and volition".

Dictionary.com provides a more detailed definition that covers a multitude of aspects of consciousness, namely, consciousness is:

1. The state of being conscious: awareness of one's own existence, sensations, thoughts, surrounding, etc.
2. The thoughts and feelings, collectively, of an individual or of an aggregate of people.
3. Full activity of the mind: as in waking life: to regain consciousness.
4. Awareness of something for what it is; internal knowledge: consciousness of wrong doing.
5. Concern, interest or acute awareness: class consciousness.
6. The mental activity of which a person is aware as contrasted with unconscious mental processes.
7. Philosophy: the mind or the mental faculties as characterized by thought, feelings and volition.

Some particular views of consciousness are the following:

Baars (1988) says that "consciousness is accomplished by a distributed society of specialists which is equipped with working memory, called a *"global workspace"*, whose contents can be broadcast to the system as a whole" and that "only one consistent content can be dominant at any given moment". The content of the memory is decided by the consciousness [1]. This account of consciousness presupposes a neural network approach in which there is a competition among neural coalitions that involve both frontal and sensory areas, the winning coalition being conscious [2].

Rosenthal (2000) argues that an *experience* (mental state) is *"phenomenally conscious"* only in virtue of a higher-order state which is a thought or perception representing that experience. This approach is called *higher-order theory* (**HOT**). In HOT phenomenally conscious states are the states which possess fine-grained intentional contents of which the subject is *aware*, being the target or potential target of some kind of higher-order representation [3]. For example, a conscious experience of yellow consists in a representation of yellow in the visual system accompanied by a thought in the same subject to the effect that the subject has the experience of yellow.

Dennet (2001) points out that there is no single central place where conscious experience occurs, but there are "various places and times in the brain". When "content-fixation" occurs in one of these places its effects may propagate, and so it leads to the expression of one of these sentences which generate the story with central character one's self. For Dennet consciousness is a serial representation of the brains underlying parallelism [4].

Some particular scientific views of consciousness are the following:

Greenfield (2000) proposes the concept of a *"continuum of consciousness"* and argues that "Consciousness is a dynamic process and it changes with the development of brain. The macro level does not contain a consciousness center, and at the micro-level there are committed neurons or genes dedicated to consciousness" [5].

Hawkins (2004) says that consciousness is a combination of self-awareness and feelings with sensations, but not related to sensory input (what is called, *qualia*). He argues that memory and prediction play important roles in the creation of consciousness, no matter how it is defined. Once the memory is erased conscious experience disappears [6].

Tononi (1998) suggests a functionalist theory of consciousness, called the *"integrated information consciousness theory"* according to which the levels of consciousness of an agent at a time depend on how many possible states it has at the time, and how tightly integrated are its states are [7]. This view of consciousness implies that there would be a loss of consciousness in a *seizure* in which the number of possible states is razed. The information associated with the occurrence of a conscious state is information from the perspective of an integrated system. When one experiences a particular conscious state, that conscious state is experienced as an *integrated whole*, and cannot be decomposed into independent components, i.e., components that are experienced are measured using Shannon's information theory [8, 9].

Searle (1992) expresses the view that "studying the brain without studying consciousness would be like studying the stomach without studying digestion, or studying genetics without studying the inheritance traits" [10].

Sloman (1991) has the view that "it may be pointless trying to define consciousness, its evolution or function as they may have different interpretations similar to other big words like perception, learning, knowledge, attention, etc." [1].

Place (1956) and **Smart** (1959) founded the *biological theory of consciousness,* according to which consciousness is some kind of biological state of the brain [12, 13]. For example, the experience of seeing green is interpreted as a sort of activated brain state. The stimuli create a biological reaction in the brain which leads to the experience of green. The question is "what makes such a representational content phenomenally conscious? One view is that active connections between cortical activations and the top of the brain stem constitute the so-called *"thalamic switch"*, which however is an **ON** switch rather than an **OFF** switch. This is supported by evidence that the transition from the unconscious (negative) state to the minimally conscious state involves these connections, whereas corticothalamic connections are disabled as a result of the overall decrease in cortical metabolism [14, 15].

Block (2009) argues that the biological theory handles better the *"explanatory gap"* (i.e., the fact that we have no idea why the neural basis of an experience is the neural basis of that experience rather than another experience or no experience at all) than do the *global space* or higher *order theories* [16]. He says that "according to the biological view, both the global workspace and higher-order thought views leave out too many details of the actual working of the brain to be adequate theories of consciousness". Also, "according to the biological account, global broad-casting and higher-order thought are what consciousness *does* rather than what consciousness **is**". On the other hand the global workspace and higher order accounts link consciousness to the ability to report it more tightly than does the biological view.

Pitrat (2009) distinguishes four meanings of consciousness, namely [17]:

- Consciousness as perceiving the world.
- Consciousness as being awake.
- Consciousness as being unique.
- Consciousness a observing its reasoning.

Consciousness as perceiving the world This is what is called the *"phenomenological consciousness"*, and is based on the sensations that our brain receives from the world, and on the perceptions obtained from our processing of these sensations. On the basis of this information we decide to perform some actions. The representation of the external world is what is typically called *"awareness"*.

Consciousness as being awake This is the state in which our brain is *awake* (we know that it is running, we perceive information from the external world, and we act on it). Being awake implies that we have not lost consciousness because of deep sleep or coma or anesthesia, etc.

Consciousness as being unique This is the feeling of existing, of being one's self and unique, different from other beings. It is what is more commonly called *"self-awareness"*. A typical example of this kind of consciousness *(self-awareness)* is when a person sees him/her self in the mirror and recognizes that the image in the mirror is the mirror of his/her self *(mirror consciousness test)*.

Figure 7.1 Seven levels of consciousness.

Source: https://www.withfriendship.com/images/i/43519/levels-of-consciousness.gif

Consciousness as observing its reasoning   This is what is called *"access consciousness"*. Our brain has the capability to access information for reasoning and for controlling our performance according to the results of the reasoning. Using these results we can check the correctness of our reasoning and make a verbal report to other persons who will more readily accept or employ our conclusions. If a person realizes that has started in a wrong way, he/she stops immediately and attempts to find a better way. Figure 7.1 illustrates the four kinds of consciousness classified according to human needs, namely: spiritual, mental, emotional, and physical. These consciousness kinds involve seven levels of consciousness according to human motivations.

7.3 Machine – Robot Consciousness

The issue whether or not human-like conscious machines and robots are possible is still an open question. However, many artificial intelligence/artificial consciousness scientists believe that many different cognitive skills possessed by humans can be implemented in machines (computers and robots). Therefore the field of machine consciousness (or synthetic consciousness) is typically considered as a subfield of artificial intelligence and cognitive robotics that aims to specify what functions would be synthesized in order for an artificial being to be characterized as conscious. Thus, machine consciousness has to address the following fundamental questions:

- What is the proper combination of cognitive capabilities?
- What cognitive capabilities are absolutely needed for consciousness?
- How exactly should be synthesized and implemented?
- Which capabilities should be implemented first?

Of course, to address the above questions we must set aside the philosophical implications by phenomenological dynamic, and complexity issues of multiple consciousness processes, and work merely on the basis of the functional issues of consciousness. For example, a machine or robot needs to be *aware* of the world. This can be achieved via its sensors, while the required processing of the sensory information can be performed by its programs. A machine can be thought as being *awake* when it is running (except when it is running in an infinite loop). A machine can have *access consciousness* more perfectly than a human, since it may have more complete access to its processes while it is working.

However a machine is very difficult to have a physical image and be able to recognize itself in a mirror. Moreover a machine does not know that it exists and that there are other beings. It only receives our orders and try to fulfill them, without knowing that we exist and can modify our commands. Self-awareness is a very hard function for implementation in a machine. Takeno has deeply studied self-awareness and asserted that he has developed a robot capable of discriminating between a self-image in a mirror and any other being that has an identical image to it [18]. He constructed a mirror image cognition experiment using a robot equipped with the **MoNAD** system.

Although most researchers have different views of machine consciousness, they agree that the envelope concept *"machine consciousness"* needs to be decomposed in functional terms. They agree that machine consciousness will be very much a function of embodied form of intelligence that machines will possess, i.e., consciousness is an emerging property of machine development and its final form will depend on cognition, perception, memory, learning, motor capabilities, thoughts, emotions, motivations, plans, etc.

Some representative views in the debate about machine consciousness are the following.

Buttazzo (2001) argues that despite our present technology's ability to simulate autonomy, "the most common objection to granting electronic-driven computers self-conscious status is the perception that, working in fully automated mode, they cannot exhibit creativity, emotions or free will" [19].

Alexander (1996) who has studied deeply artificial neural networks points out that the problem of artificial consciousness becomes more intriguing because neural networks replicate the basic electrical behavior of the brain, and provide the proper support for realizing a processing mechanism similar to that adopted in the brain. However, although the principles of creating conscious machines already exist it would take forty year to train such a neural network-based machine to understand language [20]. Of course the assumption that the brain is a neural state machine needs to be proven [21].

Chalmers (1994) provides explicit arguments for the plausibility of machine consciousness. He argues that "Computers perform computations. Computations can capture other systems' abstract causal organization. Mental properties are nothing over and above abstract causal organization. Therefore, computers that run the right kind of computations will instantiate mental properties" [22]. He justifies the *"organizational invariance"* of the mental properties (i.e., that they are nothing over and above abstract causal organization) by the argument that mental properties are of two types:

- *Psychological properties* ((e.g., belief and perception) which are "characterized by their causal role" within an overall causal system.

- *Phenomenological properties* which are not *prima facie* definable in terms of their causal roles.

For the psychological properties he accepted the view that systems with the same topology will share their psychological properties. For the establishment of the view that phenomenological properties are amenable to individualization he provided his *"dancing qualia argument"* [23]. (See also [24]). Chalmers' assumption that all mental properties are sufficiently captured by abstract causal organization has been challenged and criticized by many researchers.

Thaler (2009) developed the so-called *creative machine* (Device for the Autonomous Generation of Useful information), proposing a neural cascade as a canonical model of consciousness, where the first network implements a *stream of consciousness* and the second network develops an *attitude* about the cognitive turnover within the first network (i.e., the subjective feel of consciousness). This creative machine was the product of his *"confabulation theory"*, in which all aspects of human (and animal) cognition are modeled in terms of confabulation (or false memories) generation [25]. He patented a series of artificial neural network paradigms that are essential for creating artificial brains. These will be discussed in Chapter 9.

Damasio (1999) argues that a living being achieves self-awareness when its brain models its body, the outside world, and their interactions. This is just a moment-to-moment self-awareness. If the brain has memory and remembers the past then the being is aware of itself as something which has a history. According to him the above basic self-awareness arises as follows [26]:

- The body uses a myriad of internal sensors for sensing chemicals in the blood, the blood pressure, the heart rate, etc. So a machine that has internal sensors and the right software from them can also form a model of the body at each instant.
- The above body is embedded in a larger model (which is called *Damasio's second-order model*). In this larger body, there is the body, there is the outside world as sensed by the body's external senses, and also there are interactions between this outside world and the body, again as sensed by them.

Thus, at each instant, the brain has two simultaneous views, namely: a view *"from the outside"* (e.g., the brain, as observer, sees from its second-order model the body which has a hand, and the hand extending to pick-up a tea cup) and a view *"from the inside"* (e.g., its eyes see the cup and, *ouch,* its skin gets burned by the hot cup). As the hand quickly pulls back, both models simultaneously report the same thing. From this, human consciousness arises and Damasio points out that there is no reason that machine consciousness cannot arise in a similar manner [26].

7.4 Requirements for Machine – Robot Consciousness

Conscious robots will need to be embedded in the real world with significant input/output capabilities and learning from humans and experience capabilities. Robot learning from humans (and interaction with them) needs care and dedication on the part of the human, especially in the area of social or companion robots.

Over the years several sets of requirements for machine-robot consciousness were proposed depending of the views held by the respective researchers. Some of them are the following.

Baars [1] Key functions for conscious machines include:

- Definition and context setting.
- Adaptation and learning.
- Editing.
- Flagging and debugging.
- Recruiting and control.
- Prioritizing and access-control.
- Decision making.
- Executive function.
- Analogy-forming function.
- Metacognitive and self-monitoring function.
- Auto-programming and self-maintenance function.

McCarthy [27] Robot self-consciousness needs at least the following capabilities:

- Observing its physical body.
- Observing that it does or doesn't know the value of a certain term.
- Keeping a journal of physical and intellectual events.
- Observing its goal structure and forming sentences about it.
- Observing its intentions.
- Observing how it arrived at its current beliefs.
- Representing or observing auxiliary information and rising sentences about it.

- Regarding its entire mental state up to the present as an object (i.e., as a context).
- Knowing what goals it can currently achieve and what its choices are for action.

It is noted that in the above list something is considered as observable when the robot causes a sentence (and possibly other data structures) that gives the result of the operation to be exhibited in the robot's consciousness.

Aleksander [28] The principles for machine consciousness are:

- The brain is a state machine.
- Inner neuron partitioning.
- Conscious/unconscious states.
- Perceptual learning and memory.
- Prediction.
- The awareness of self.
- Representation of meaning.
- Learning utterances.
- Learning language.
- Will.
- Instinct.
- Emotion.

Block [29] Consciousness is distinguished in the following four-not necessarily different-types:

- *Phenomenal consciousness* (what makes a state phenomenally conscious is that there is something – it is like – to be in that state).
- *Access consciousness* (for an experience to be called access conscious, the content of the experience must be available to the direct "rational" use of the subject).
- *Monitoring consciousness* (a higher-order consciousness, e.g. an internal scanning-like mechanism, or a second-order mental state which accompanies first-order conscious mental states).
- *Self-consciousness* (ability to conceptualize a self; all phenomenal consciousness states have a perspective of the subject, but the knowledge of one's self – conceptualization of one's self – is required for self-consciousness).

Lycan [30] The following problems of consciousness correspond to the various features of consciousness:

- **Problem 1:** Difference between conscious and unconscious states (how and on what basis some mental states are conscious while others are not).
- **Problem 2:** Explanation of introspection (one has knowledge of his/her conscious experiences' content via introspection which is not readily available to any other person).
- **Problem 3:** The qualia concept (the qualitative properties of an experience, often called "raw feels of an experience").
- **Problem 4:** Explanation of the homogeneity of conscious experience (i.e., experiences have smoothness and contiguousness which do not appear to occur in the external physical world).
- **Problem 5:** Intrinsic perspectiveness of conscious experiences (i.e., first-person perspective of the conscious experiences).
- **Problem 6:** The explanatory gap problem (i.e., the problem of explaining the connection between the physical neurological facts and the subjective properties of a conscious experience).
- **Problem 7:** The funny facts problem (i.e., the problem of explaining any facts that are nonphysical).
- **Problem 8:** Ineffability of what is like (One often cannot express in words what it is like to have a particular sensation).

Clearly, although at first glance the above sets of principles, features and requirements seem different, actually they are similar, although some overlapping or redundancy occurs. In general, it is thought that the goal of creating a conscious machine/robot will have been achieved if one can embed to it all the above features, although the philosophical doubts as to whether the robot is "really" conscious might remain. From a practical point of view roboticists follow an *"engineering"* approach trying to create "a system that does the kinds of things that, when done by a human require consciousness. It is sufficient for them to develop a robot that behaves in such a way that, if it were a living being, we should assume that it is aware.

A minimal set of features/requirements for a robot with acceptable consciousness that can be drawn from the above sets might be the following [31]:

- *Difference* (Ability to distinguish conscious and unconscious mental states).
- *Phenomenal* (Ability to unite different modalities of perception in a single experience).
- *Subjectivity* (All conscious experiences belong to a single robot, i.e., they are from the point of view of the robot).

- *Availability* (The content of conscious mental states must be readily available for use by the entire system).
- *Explicit and direct knowledge* (Ability to monitor consciousness and know directly and explicitly the content of conscious mental states).
- *Qualia* (Possession of raw feels or sensing data that provide the *"qualitative features"* of an experience).

A more technical list of conscious robot requirements, inspired by Alexander's principles, is provided in [32]. This list is the following:

- Autonomy.
- Adaptability/advanced learning capabilities.
- Emotion/affect.
- Responsibility (or being something to which we are responsible).
- Intelligence.
- Authenticity (own world view and goals).
- Ability to integrate information from different sources/modalities.
- Vivid/meaningful sensation/perception.
- Ability to act in the world.
- Ability to simulate/imagine/plan.
- Ability to represent its own states.
- Attention capabilities.
- A belief that it is conscious/an ability to give phenomenological reports.

In the literature, a number of computational consciousness models and architectures have been proposed which possess certain subsets of the above lists and exhibit several types and levels of consciousness. These models are based on symbolic methods and/or neural network methods. A few of them are briefly outlined in the next section. More models/architectures will be considered in Chapter 9.

7.5 Machine Consciousness Implementation Models

The following machine consciousness models will be considered:

- *IBCA:* Integrated Biologically-based Cognitive Architecture.
- *CLARION:* Connectionist Learning with Adaptive Rule Induction ON-line.
- *ACT-R:* Adaptive Control of Thought-Rational.
- *CERA: Conscious and Emotional Reasoning Architecture.*
- *MANN: Modular Artificial Neural Network architecture.*

7.5.1 The IBCA Model

IBCA employs a model of the brain working memory and epitomizes the automatic and distributed concepts of information processing performed in it [33, 34]. Controlled processing is an emergent property of the dynamic interactions of multiple brain systems. IBCA emphasizes the influence of three regions (systems) in the brain, namely *prefrontal cortex* (PFC), *hippocampus* (HCMP) and *perceptual and motor cortex* (PMC).

PFC is dedicated to active maintenance of internal contextual information which is dynamically updated and self-regulated so as to bias ongoing processing according to maintained information (e.g., goals, instructions, etc.).

HCMP is dedicated to rapid learning of arbitrary information, which can be recalled during controlled processing.

PMC performs slow, long-term learning which accumulates knowledge and skills.

Table 7.1 shows a classification of the three systems according to the following features [33].

- *Function* (which specifies the function optimized by this system).
- *Internal relation* (which indicates how representations within each system are interrelated).
- *External relation* (which specifies how representations relate to other systems).
- *Act capacity* (which shows how many representations can be active at any given time).
- *Learn rate* (which indicates the characteristic rate of learning).

A salient feature of IBCA is that it integrates biological and computational principles and involves a large repertoire of specialization in different brain systems (e.g., robust active maintenance, fast vs. slow learning). It can perform generalization, parallelism and flexibility, and in addition it has

Table 7.1 Parametrization of PMC, PFC, and HCMP [33]

System	Function	Internal Relation	External Relation	Act Capacity	Learning Rate
PMC	Inference, Processing	Distributed, Overlapping	Embedded	Many	Slow
PFC	Maintenance, Control	Isolated, Combinatorial	Global	Few	Slow
HCMP	Rapid learning	Separated, Conjunctive	Context sensitive	One	Fast

complementary learning capacities for resolving problems with knowledge consolidation, or transfer of short-term into long-term memory.

As shown in Table 7.1, PMC optimizes knowledge-dependent *inference* capabilities, and slow integrative learning, i.e., integrating over individual learning episodes. PFC optimizes *active memory* via restricted recurrent excitatory connectivity and an active gating mechanism. HCMP optimizes fast learning of information in weight-based memories which allows the binding of elements of a new association (including representations in PFC and PMC). According to [33], although there are many important specialized brain models, the proposed three brain systems/areas (PFC, HCMP, PMC) provide central and crucial explanations to working memory function.

Figure 7.2 provides a diagram of the key properties of these principal brain systems. PMC representations are distributed and embedded in specialized processing systems, PFC representations are isolated from each other, and combinatorial (i.e., there are separate active units that represent the values of each feature). HCMP representations are sparse, conjunctive and separated but still distributed.

In Figure 7.2, active representations are shown in grey, distributed representations are shown by highly overlapping circles, isolated representations by non-overlapping circles, and separated representations in between. Connections of active units are shown by solid lines, and connections of non-active units by dashed lines. Learning is performed by the **LEABRA** *learning framework* shown in Figure 7.3(a, b). This framework combines

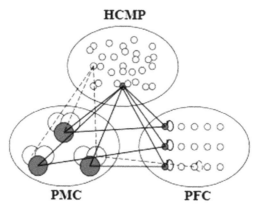

Figure 7.2 Pictorial representations of the three principal brain systems HCMP, PMC and PFC.

Source: http://cognitrn.psych.indiana.edu/busey/q551/PDFs/Reilly,%20Braver,%20Cohen%20-%20A%20Biologically-Based%20Computational%20Model%20of%20Working%20Memory.pdf

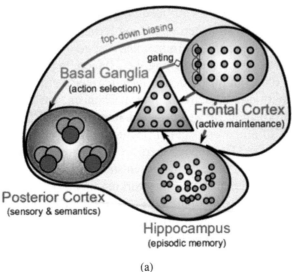

(a)

(b)

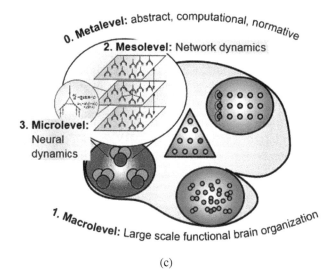

(c)

Figure 7.3 Leabra cognitive architecture/mechanisms (a) Macro-level structure, (b) Micro-level structure, (c) Integated 4-level architecture (metalevel, macrolevel, mesolevel, microlevel).

Source: http://www.psych.colorado.edu/~oreilly/papers/oreillyhazyherdip.pdf

error-driven learning (weight adjustment) and Hebbian learning with inhibitory competition dynamics. The integration of these two learning mechanisms has several advantages over many other algorithms, and assures accurate results.

7.5.2 The CLARION Model

CLARION (**C**onnectionist **L**earning **A**daptive **R**ule **I**nduction **ON**-line) is a hybrid rule-based and neural network (connectionist) system that possesses a two-level representation explaining the distinction between conscious and unconscious mental processes. It recognizes both explicit and implicit knowledge, and has been successful in working with a variety of psychological data. CLARION was used for simulation of several well-known skill learning tasks ranging from simple reactive skills to cognitive skills. It was developed by Ron Sun research team [35–37].

The explicit knowledge is represented by semantic rules and chunks (i.e., collections of dimension/value or property/value pairs). The state of the world is represented by a set of property/value pairs. CLARION uses a working memory (that stores information/human thought processes in the short term) and also goals which, among others, involve property/value pairs. The rules use chunks and scan the property/value pairs in the world, the working memory, and the goals. A rule is triggered when all property/value pairs in the premise part and the rule are present, in which case the property/value pairs contained in the conclusion (action) part of the rule enter the working memory. CLARION's working memory resembles the human short-term memory.

The subsymbolic (implicit) knowledge is represented by a neural network with back propagation-based training. The CLARION model involves two levels: the top level where explicit (conscious) knowledge is handled and the bottom level where the implicit (subconscious) knowledge is treated. The principal subsystems of CLARION are:

- Action-centered subsystem (**ACS**)
- Non-action-centered subsystem (**NACS**)
- Motivation subsystem (**MS**)
- Meta-cognitive subsystem (**MCS**)

Each subsystem possesses top-level and bottom-level modules to handle explicit and implicit knowledge, correspondingly. Property/value pairs that result in the outputs of the rules or back propagation networks are stored in the working memory and degrade or decay with time. Thus, the working memory is changing over time, and its items are available to both the top and low levels of the system. To stack and handle the hierarchical nature of goals and sub-goals, CLARION implements a goal structure as an individual component. The general organizational architecture of CLARION is depicted in Figure 7.4 [37].

CLARION meets the following machine consciousness requirements:

- *Distinction ability* between conscious and unconscious processes with each 2-level structure. The top level for the conscious and the bottom level for the unconscious mental states.
- *Phenomenal consciousness* because there is a representational difference between the top and bottom level processes.
- *Qualia:* Again by the separate representation of top and bottom level processes (although such a division may not be sufficient in all cases to model qualia).

The tasks that can be performed by CLARION include **SRT** (serial reaction time) tasks, **PC** (process control) tasks, the **AA** (alphabetic arithmetic) task, and the **TOH** (Tower of Hanoi) task, etc. The SRT, AGL and PC tasks are standard implicit learning tasks associated with the processes of consciousness.

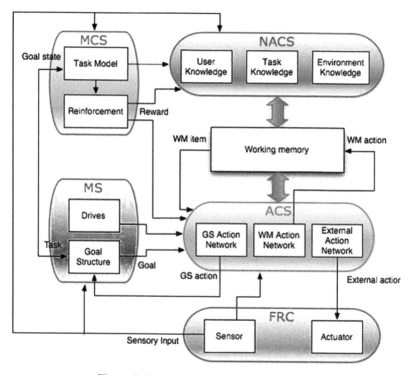

Figure 7.4   The CLARION architecture [37].

7.5.3 The ACT-R Model

The *ACT-R model* is an open architecture that can be used to carry out several tasks [38, 39]. Although ACT-R is not explicitly attempting to model consciousness *per se*, it exhibits features of consciousness that are present in other computational models designed specifically to address consciousness. For example, in ACT-R the distinction of conscious and unconscious processes is achieved by having some information in the modules, and some information placed in the buffer memory. Availability of conscious information to the other processes of the system is assured by the fact that the production system of ACT-R has access to all the information that exists in the buffers. In other words, the information, once placed in a buffer memory, becomes available to the entire system. The ACT-R model has the general structure shown in Figure 7.5 [39].

Many of the processes in ACT-R are performed in parallel, although some other processes are carried out serially. The system involves a set of modules

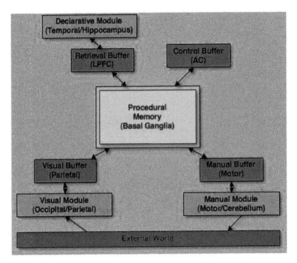

Figure 7.5   The ACT-R architecture [39].

Source: http://www.ai.rug.nl/~niels/images/actr.jpg

(e.g., vision module, motor module, intention module, declarative module, etc.). Each module contains a buffer which selectively holds the information required for the task at hand. These modules/buffers interact with the external world. Their mutual interaction is performed via production rules which scan the buffers. The rules that fire generate results that are also places in the buffers. Therefore, the contents of the buffers change dynamically as new inputs are received and the rules are applied. Each module places information concerning its operations into its buffer. The production system monitors the modules behavior patterns and fire rules according to the observed patterns. This ensures a strong coupling among the modules, which reflects to the integrated nature of ACT-R that does not appear to satisfy the explicit/direct knowledge, the qualia, and the subjectivity requirements of consciousness.

A system that followed the ACT-R architecture attempting to be a general-purpose intelligent system is the **SS-RICS: S**ymbolic and **S**ub-symbolic **R**obotics **I**ntelligent **C**ontrol **S**ystem. This system allows, in addition to the symbolic/production rules representation, sub-symbolic neural network representation for processing sensor data. The hybrid symbolic/sub-symbolic structure of SS-RICS has the generic form shown in Figure 7.6 [40].

Hybrid symbolic/sub-symbolic models are believed to have greater potential to model consciousness. Rule-based (symbolic) systems alone cannot provide good model of consciousness, since this would require humans to program the rules (which cannot be done for the required billions of rules).

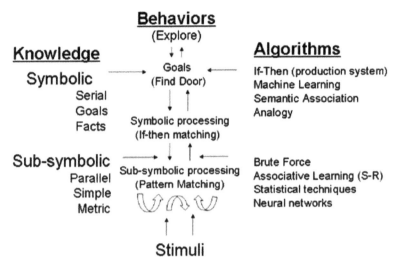

Figure 7.6   The SS-RICS hybrid computation structure.

Source: http://ss-rics.org/Publications/A%20Hybrid%20Symbolic%20and%20Sub-Symbolic%20Intellig ent%20System%20for%20Mobile%20Robots.pdf

The machines will need to rely on hybrid systems with learning and emergent behavior.

7.5.4 The CERA Model

CERA is a software framework which allows several reasoning/consciousness components to be integrated and tested in a single autonomous system [41]. The CERA alternative components were implemented using the object oriented design methodology. CERA involves the following modules:

- Attention module.
- Status assessment module.
- Global search module.
- Preconscious management module.
- Contextualization module.
- Sensory prediction module.
- Memory management module.
- Self-coordination module.

These modules are deemed to be, at minimum, necessary for any implementation of machine consciousness. In a complete model of mind these concepts must be integrated with sensory preprocessors, planning system, belief structure, memory system, and so on. CERA integrates all the above modules in a whole operational system. CERA involves three nested hierarchical layers as shown in Figure 7.7.

The inner layer, called *CERA CORE Layer* encloses the *reasoning consciousness model* (**RCM**) which is based on Baars global workspace theory. Thus, there are some similarities of CERA with the IDA system. The CERA core layer uses an abstract knowledge class to make the high-level RCM processes definition representation-independent. Thus, actually the CERA core *per se* cannot be instantiated. This is done at the next level, the *instantiation layer* where the cognitive model is built. On top of the instantiation layer, the physical layer is required to adapt the cognitive components to the actual sensorimotor machinery of the conscious autonomous robot. The simulation and controller testing was done on the Khepera platform. In this way an instantiation of CERA, called K-CERA (Khepera CERA) was obtained which is described in [41]. In K-CERA the foundation reasoning classes for the specific domain of unknown environment exploration, of Khepera, were integrated with the rest of possible cognitive components of a model of the mind. The reasoning consciousness model within CERA architecture is shown in Figure 7.8.

7.5.5 The MANN Model

This is a cognitive model for *self-consciousness* and *self-representation* [42]. It is based on the concept of *"neural correlates for consciousness"* (**NCC**), which are defined to be the neural systems and properties of these systems that are associated with conscious mental states, or, better, neural system (S) plus a certain state (NS) of this system, which together are correlated with a certain state of consciousness (C) [43]. The model uses artificial neural networks for abstract self-representation, representation of other individuals, and function evaluation, i.e., for representing the process of using

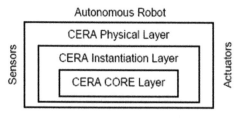

Figure 7.7 The CERA Layered structure.

Source: http://www.conscious-robots.com/raul/images/stories/cera1.gif

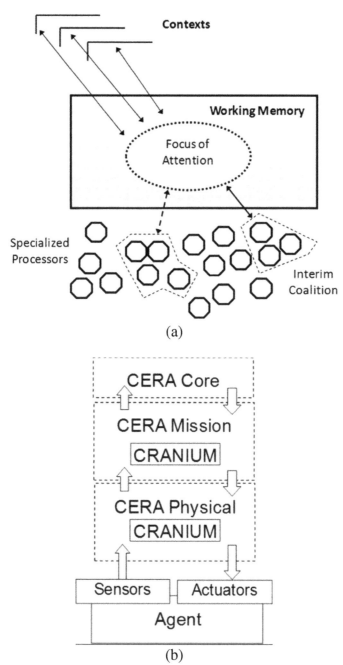

Figure 7.8 The reasoning consciousness structure within CERA [41]. (a) CERA Cognitive architecture (global workspace model), (b) CERA CRANIUM hierarchical structure.

Source: aigamedev.com/open/article/conscious-bot

self-information to predict and decide future actions. In particular, the neural networks used are multilayer perceptrons, although any other type of neural networks, with proven universal approximation capability, can be used. The basic information processing component is called *holon* (from the Greek word όλον (holon) which means whole/entire). Holons are independent elements that behave autonomously and are self-organizing, recursive, and cooperative. An holon contains information processes and possibly physical processes, it behaves autonomously, but at the same time, as part, of bigger whole.

Consciousness is treated as an holonic system including *basic features* (reactivity, adaptability, associative memory, learning ability, and optimization ability), *intermediate features* (composition or interaction of basic features, namely: abstraction, prediction, generalization, inference, emotion, motivation and imagination) and *advanced features* (free will, moral judgment and self-consciousness). The MANN model is concerned with self-consciousness, which includes an abstract and dynamic self-representation, a mechanism for using other cognitive features to evaluate this representation for reasoning and anticipation.

The cognitive model of self-consciousness was implemented and tested using *modular artificial neural networks* (**MANN**) through intelligent virtual holons, immersed in a medium, i.e., in an environment that can transmit signals or phenomena. The cognitive features (holon features) are represented by a feature vector with a different level of development in each vector variable, and there is also the holon's perception itself. As stated in [42], an artificial neural network is a good model for self-consciousness and self-representation since it has the following capabilities:

- Provides an abstract representation needed for personality of the artificial being (machine, robot, etc.).
- Senses the environment continuously.
- Changes the artificial being's representation adaptively on the basis of the environment senses.
- Uses self-information in connection with other cognitive features, such as inference, anticipation and optimization via a learning process.

The structure of the MANN model is shown in Figure 7.9 (a–c) [42]:

Each holon is able to recognize other being's capabilities and views itself (self-representation) as shown in Figure 7.9a–c. The self-representation information is used by the holons central process in order to evaluate its possibilities compared with other beings. Each holon has a certain number

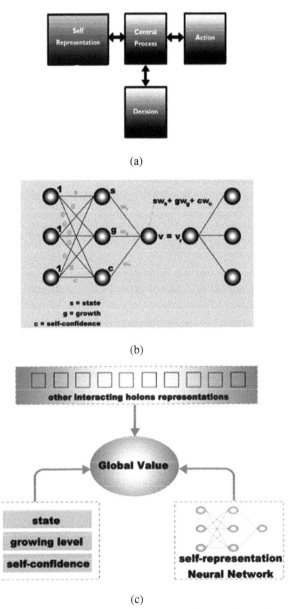

(a)

(b)

(c)

Figure 7.9 (a) MANN cognitive architecture, (b) neural network topology (s = state, g = growth, c = self-confidence), (c) learning process of the main ANN module. Courtesy of Milton Martinez Luaces [42].

Source: http://www.naun.org/main/NAUN/mcs/mcs-88.pdf

of ANNs (hence it is a modular-ANN: MANN). The topology of each ANN is as shown in Figure 7.9b where the impact of each feature is represented by the weights that connect the hidden and output layer neurons. As a result of the holon's ANN processing, an output is provided which represents the global value (gv) of each holon from its own view point (self-representation). Learning (i.e., the adjustment of the weights) is performed by typical back-propagation [44]. The main ANN module learning process is pictorially illustrated in Figure 7.9c, and involves the holons' features (s, g, c), and the global value $gv = w_s s + w_g \cdot g + w_c \cdot c$, where w_s, w_g and w_c are corresponding weights (connection strengths). The details of the implementation of the MANN consciousness model are provided in [42].

In [45] a biologically inspired machine consciousness model is presented which is driven by competing motivations, goals, and attention switching. This model consists of three principal interlinked functional subsystems, namely:

- *Sensory motor subsystem* (which involves the motor and sensor units, data encoders/decoders, sensor and motor processors, emotions, rewards, and sub-cortial processing).
- *Episodic memory* and learning subsystem (which contains the episodic memory, and queuing/organization of the episodes).
- *Central executive subsystem* (which involves modules for thinking and planning, attention switching action monitoring, and motivation/goal processor).

Finally, in [46] an emotional consciousness *(EMOCON)* machine model is provided which includes components for differentiation, integration, intensity, valence, and change. This model was produced using a unified theory of emotional consciousness that integrates a variety of components, including somatic representation, cognitive appraisal, neural- affective decision making, and working memory.

7.6 Machine Consciousness Testing and Measurement: The Con Scale

Testing whether a machine is conscious and at what level is a difficult task, more complex than testing machine intelligence. Consciousness is a grand integration of a large number of cognitive processes. In humans consciousness involves several levels which are typically determined by neurological clinical diagnosis methods and described by various scales, such as the *Glascow*

Coma Scale (**GCS**) or the *Simplified Motor Score* (**SMS**) which, however, do not cover the wide range of consciousness. Consciousness has also been measured from a psychological point of view which leads to several scales that are focused in particular issues of consciousness (e.g., the Private Self-Consciousness Scale). These neurological and psychological measures of consciousness cannot be directly applied or adapted to robots and other artificial beings.

The most well-known method for testing machine intelligence, the Turing test, cannot be adapted to machine consciousness, because when interpreted as "merely observational" is not compatible with the principles of theory dependence of observations of the philosophy of science. Also, it cannot provide a measurement of the level of consciousness of beings not reaching the human-like level, and more seriously, even at that level mere observation of external behavior is not sufficient. A good approach for testing and measuring machine consciousness is to combine *behavioral assessment* and *inner inspection* (e.g., looking for hardware or software architectural hallmarks, or calculating the capability information integration of the system as suggested by Tononi [8], or looking for the presence of axiomatic features as suggested by Aleksander and Dunmail [20, 47].

As discussed in Section 7.2 the "information integration" (unity of conscious experience) indicates the level of phenomenal consciousness. But an effective measure of machine consciousness should combine both information integration and information differentiation (ability to discriminate conscious experiences among several consciousness scenes), and assess the balance between integration and differentiation.

A comprehensive survey of machine consciousness measures is provided in [48]. Here, we will briefly outline the **ConScale**, a bio-inspired scale for measuring consciousness of artificial beings [49]. ConScale provides a taxonomy (defined as an ordered list) that classifies artificial beings according to their level of consciousness (in contrast to unitary consciousness tests that are not designed as gradual scales [48]). Also, ConScale can be used as a roadmap for the design of advanced implementation of machine consciousness, including a variety of artificial agents (physically-situated beings like robotic beings, and software beings like web agents). The ConScale consciousness levels in incremental progression, where each higher level subsumes lower ones are the following:

Level −1: Disembodied This is the initial reference level where no characteristic cognitive skills are involved. Its biological analog is an animo acid as part of a protein.

Level 0: Isolated   Here, there is no autonomous processing or sensorimotor machinery. A valid biological analog of this level is an isolated chromosome.

Level 1: Decontrolled   Here, sensors and actuators are present but without any working relation between them, which means that no cognitive skills are involved. Dead bacteria is a biological analog of this level.

Level 2: Reactive   Here, sensors and actuators are functionally cooperating leading to fixed reactive responses, and the only existing cognitive skill is primitive situatedness based on reflexes. A virus is a biological analog of this level.

Level 3: Adaptive   Here, the agent's action is a dynamic function of both memory and current sensory information. The skills at this level are the ability to learn new reflexes and use proprioceptive sensing for orientation and positioning actions. An illustrative biological analog of this level could be the earthworm.

Level 4: Attentional   Here, the agent's behavior is biased by attention (via which specific contents are selected out of the repertory available from sensors and memory). The cognitive ability at this level permits the agent to develop attach and escape behaviors. A plausible biological analog of level 4 is fish.

Level 5: Executive   Here, agents can interleave multiple goals as different working sets are represented in memory. Cognitive skills include shifting and basic emotional learning. Proper biological analogs are quadruped mammals.

Level 6: Emotional   Cognitive skills at this level are supported by the *"Theory of Mind"* (**ToM**) skills, namely: (1) "I know", (2) "I know I know", (3) "I know you know", and (4) "I know you know I know". At this level, feelings appear as representations of organism changes due to an emotion. The agent generalizes the learned experience to its general behavior. Emotions are assigned to the self, leading to a self-evaluation that produces the sense of "I know". Monkes are the biological analog of this level.

Level 7: Self-conscious   Here, ToM stage 2 skill "I know I know" represents the agent's self-consciousness. A model of the agent's self makes possible self-recognition, and learning enables the prediction of the future. The agent can plan about itself, and learn if the plan was efficient or not, and also can learn to use tools. Human babies aged 18 months are the analog of this level.

Level 8: Empathic   Here, the agent possesses an enhanced internal model involving the self and also being able to model others as selves. Thus the agent has inter-subjectivity skill i.e., "I know you know" that enables it to exhibit social behaviors. Biological analogs at this level are chimpanzees.

Level 9: Social   Here, the internal model of other selves is enriched with ToM stage 4, i.e., "I know you know I know", and so the agent can exhibit social intelligence which includes social behaviors (e.g., leadership, knavery, lying, etc.). This level is also characterized by reporting and linguistic capabilities. Artificial beings with social skills can develop 4-year old human behaviors.

Level 10: Human-Like   Obviously, the biological analog of this consciousness level is the adult human. Skills include accurate verbal reporting, use of tools for learning, extension of knowledge via external media (e.g., written communication) and technological advances, and also the ability to modify the environment.

Con Scale also includes a virtual super-conscious level (Level 11) which has no exemplifying analog in the biological world. The scale is a generic tool that can be applied to real application domains by designing concrete tests. In [50] Con Scale was applied to several examples in the domain of video game agents, and is proposed for use as an engineering roadmap for designing conscious machines. It is argued that it makes to build empathic machines (level 8) in view of building in the future machines with social consciousness. (level 9).

7.7 Concluding Remarks

Scientists have made tremendous efforts to emulate consciousness by computing and other technological means. But till now they did not fully succeed to do this. Consciousness therefore seems not to be purely mechanical or chemical as materialists argue. What they have done is to simulate the body by hardware in some way, and tried to simulate the mind providing better and better approximations, which however still are far behind the desired ideal target. In this chapter we have discussed the fundamental philosophical and practical aspects of consciousness, including ontological issues for human and robot consciousness, the basic requirements for conscious machines, a number of practical computational models of approximate machine consciousness, and the issue of testing and measuring consciousness using a *consciousness scale* (**ConScale**).

Actually, human consciousness still remains a big *"unknown"*. We are very ignorant about our own consciousness, mind, and mentality. The great efforts of scientists all over the years (some of which were discussed in this chapter) offer good approximations of computational modeling of consciousness, but they are far from being integrated. Perhaps proper combinations and integrations of them could offer more powerful models. Additional questions that must be addressed in this direction are [51]:

- Can an existing person's mind be programmed, and if yes, how?
- Can a human's self be copied?
- What is an non-copiable machine like, and how does it work?
- Is the emerged consciousness copiable?
- Is there any piece of human consciousness, other than anxiety of death, that may not be emulated by a computer?

References

[1] B. Baars, A Cognitive Theory of Consciousness, Cambridge, U.K.: Cambridge University Press, 1988.

[2] C. Koch, The Quest for Consciousness: A Neurobiological Approach, Englewood, CO, U.S.A.: Roberts, 2004.

[3] D. Rosenthal, Metacognition and higher-order thoughts, Conscious Cognition, 9(2), 2000, 231–242.

[4] D. Dennet, Are we explaining consciousness yet? Cognition, 72, 2001, 221–237.

[5] S. Greenfield, The Private Life of the Brain, New York, NY, U.S.A.: John Wiley, 2000.

[6] J. Hawkins and S. Blakeslee, On Intelligence, New York, NY, U.S.A.: Henry Holt & Company, 2004.

[7] G. Tononi and G.M. Edelman, Consciousness and complexity, Science, 282 (5395), 1998, 1846–1851.

[8] G. Tononi, Information measures for conscious experience, Archives Haliennes de Biologie, 139(4), 2001, 367–371.

[9] C.E. Shannon and W. Weaver, The Mathematical Theory of Communication, Urbana, Champaign, U.S.A., University of Illinois Press, 1949.

[10] J. R. Searle, The Rediscovery of the Mind, Cambridge, MA, U.S.A.: MIT Press, 1992.

[11] A. Sloman, Developing and Brain Sciences, 14(4), 1991, 694–695.

[12] U.T. Place, Is consciousness a brain process? British J. Psychology, 47, 1956, 44–50.

[13] J.J. Smart, Sensations and brain processes, Philosophy Review, 68, 1959, 141–156.

[14] M.T. Alkire and J. Miller, General anesthesia and the neural correlates of consciousness, Progress in Brain Research, 150, 2005, 229–244.

[15] G. Tononi and K. Koch, The neural correlates of consciousness: An update, Annals NY Academy of Sciences, 1124, 2008, 239–261.

[16] N. Block, Comparing the major theories of consciousness, In: M. Gazzaniga (ed.), The Cognitive Neuroscience, IV, Cambridge, MA, U.S.A.: MIT Press, 2009.

[17] J. Pitrat, Artificial Beings: The Conscience of a Conscious Machine, Hobroken, NJ, U.S.A.: John Wiley/ISTE Ltd, 2009.

[18] J. Takeno, K. Inaba and T. Suzuki, Experiments and examination in mirror image cognition using a small robot, Proceedings IEEE Intern. Symposium on Computational Intelligence in Robotics and Automation (CIRA 2005), Espoo, Finland, June 2005, 493–498.

[19] G. Buttazzo, Artificial consciousness: Utopia or real possibility? Computer, —-?

[20] I. Aleksander, Impossible Minds: My neurons, My consciousness, London, U.K.: Imperial College Press, 1996.

[21] R.J. Wilson, Review of Impossible Minds, J. Consciousness Studies, 5(1), 1998, 115–116.

[22] D.J. Chalmers, A computational foundation for the study of cognition, J. Cognitive Science, 12, 2011, 323–357, 1994, (Available online).

[23] D.J. Chalmers, Absent Qualia, Fading Qualia, Dancing Qualia, In: T. Metzinger, Conscious Experience, Ferdinand Schoningh, 1995, www.cousc.net/papers/qualia.html

[24] T. Horgan, Functionalism, qualia, and the inverted spectrum, Philosophy and Phenomenological Research, 44, 1984, 453–469.

[25] S.L. Thaler, Home of the creative machine: The big bang of machine intelligence http://imagination-engines.com/iei_ip_overview.php

[26] A. Damasio, The Feeling of What Happens: Body and Emotion in the Making of Consciousness, New York, U.S.A.: Harcourt Brace, 1999.

[27] J. McCarthy, Making robots conscious of their mental states, AAAI Technical Report SS-95–05, 1995, (www.aaai.org).

[28] I. Aleksander, Artificial neuroconsciousness: An update, In: J. Mira and F. Sandoral (eds.), From Natural to Artificial Neural Computation, LNCS: Vol. 930, Berlin, Germany: Springer, 1995, 566–583.

[29] N. Block, On a confusion about a function of consciousness, In: N. Block (ed.), Consciousness, Functionalism and Representation (Vol. 1), Cambridge, MA, U.S.A.: MIT Press, 2007.

[30] W.C. Lycan, Plurality of consciousness, www.unc.edu/~ujanel/CogThs. html

[31] S. Gök, Modeling consciousness: A comparison of computational models, M.Sc. Thesis, Department of Cognitive Science, Middle East Technical University, Ankara, Turkey, 2009.

[32] R. Chrisley, R. Clowes and S. Torrance, Next-generation approaches to machine consciousness, Proceedings of Symposium on Next Generation Approaches to Machine Consciousness: Imagination, Development, Inter-subjectivity, and Embodiment (AISB'05), University of Hertfordshire, Hatfield, U.K., 12–15 April, 2005, 1–11.

[33] R.C. O'Reilly, T.S. Brower and J.D. Cohen, A biologically-based computational model of working memory, In: A. Miyake and P. Shah, (eds.), Models of Working Memory, Cambridge, U.K.: Cambridge University Press, 1999, 375–411.

[34] R.C. O'Reilly and Y. Mukata, Computational Explorations in Cognitive Neuroscience, Understanding of the mind by simulating the brain, Cambridge, MA, U.S.A.: MIT Press, 2000.

[35] R. Sun and X. Zhang, Top-down versus bottom-up learning in cognitive skill acquisition, Cognitive Systems Research, 5, 2004, 63–89.

[36] R. Sun, A tutorial on CLARION 5.0, 2003, www.cogsci.rpi.edu/~rsun/ sun.tutorial.pdf

[37] R. Sun, The CLARION cognitive architecture: Extending cognitive modeling to social simulation, In: R. Sun (ed.), Cognition and Multi-Agent Interaction, Cambridge, U.K.: Cambridge University Press, 2006.

[38] J.R. Anderson, ACT: A simple theory of complex cognition, American Psychologist, 51, 1996, 355–365.

[39] J.R. Anderson, D. Bothell, M.D. Byrne, S. Douglas, C. Libiere and Y. Qin, An integrated theory of the mind, Psychological Review, 111(4), 2004, 1036–1060.

[40] T. Kelley, E. Avery, L. Long, and E. Dimperio, A hybrid symbolic and sub-symbolic intelligent system for mobile robots, AIAA Paper No. 2009–1976, Info Tech. & Aerospace Conference, Seattle, WA, U.S.A., 2009.

[41] R. Moreno and A.S. de Miguel, A machine consciousness approach to autonomous mobile robotics, Proceeding First International Cognitive Robotics Workshop (AAI-06), Boston, MA, U.S.A., 2006.

[42] M.M. Luaces, C. Gayoso, J.P. Sierra and A. Rodriguez-Paton, Using modular neural networks to model self-consciousness and self-representation for artificial entities, Mathematics and Computers in Simulation, 2(2), 2008, 163–170.

[43] J. Fell, Identifying neural correlates of consciousness: The state space approach, Science Direct, 2004. www.sciencedirect.com

[44] S. Haykin, Neural Networks: A Comprehensive Foundation, Upper Saddle River, NJ, U.S.A.: Prentice-Hall, 1999.

[45] J.A. Starzyk and D.K. Prasad, A computational model of machine consciousness, International J. of Machine Consciousness, 3(2), 2011, 255–282.

[46] P. Thagard and B. Aubie, Emotional consciousness: A neural model of how cognitive appraisal and somatic perception interact to produce qualitative experience, Consciousness and Cognition, 17, 2008, 811–834.

[47] I. Alexander and B. Dunmail, Axioms and tests for the presence of minimal consciousness in agents, J. of Consciousness Studies, 10, 2003, 7–18.

[48] A. Seth, Z. Dienes, A. Cleeremans, M. Overgaard and L. Pesoa, Measuring consciousness: Relating behavioral and neuropsychological approaches, Trends in Cognitive Sciences, 12, 2008, 314–321.

[49] R. Arrabales, A. Ledezma and A. Sanchis, ConScale: A plausible test for machine consciousness? Proceedings of the Nokia Workshop on Machine Consciousness, 13th Finish Artificial Intelligence Conf., Helsinki, Finland, 2008, 49–57.

[50] R. Arrabales, A. Ledezma and A. Sanchis, Establishing a roadmap and metrics for conscious machines development, Proceedings IEEE Intern. Conf. on Cognitive Informatics (ICCI-2009), Hong Kong, June 2009, 94–101.

[51] J. Wang, On the limit of machine intelligence, International Journal of Intelligence Science, 3, 2013, 170–175.

8

Robot Conscience and Ethics

Never do anything against conscience,
even if the state demands it.
Albert Einstein

Ethics is knowing the difference between what you have
the right to do and what is right to do.
Potter Stewart

The bottom line is, robots need to be responsive
and resilient. They have to be able to protect themselves
and also smoothly transfer control to human when necessary.
David Woods

("Want responsive robots? Start with responsive humans")

8.1 Introduction

Conscience is the inner mechanism woven into our lives which tells each one of us himself what kind of entity he is as a human being, i.e., it informs an individual which is the ontological difference between *beings* and *objects* of which humans are the former. More specifically, conscience is essential in our life because it guides us in the process of choosing among alternative actions considered in our conscious reasoning. Conscious actions that can lead to the achievement of a certain goal are accepted or rejected because they agree or not agree with our conscience. Thus consciousness is a prerequisite for conscience. Consciousness and conscience both belong to the mechanisms that create the behavior of a human being. This is also true for building the behavior of robots and other artificial beings.

Ethical beings act as *delivers* or *producers* moral behavior, and as *consumers* or *recipients* of moral behavior. When a robot has certain duties or responsibilities (e.g., to protect other members of the moral community), then it is considered as a *moral producer*. When a robot is thought to have certain rights or needs that other community members ought to respect, then it may be regarded as a *moral consumer*.

The ultimate goal of machine (robot) morality/ethics is to create a machine that follows an ideal ethical principle or system in decisions about possible courses of action it could perform. If a robot is a moral producer, it cannot be automatically assumed that it is also a moral consumer, and vice versa. Features of robot moral status are: intelligence, autonomy, free will, consciousness, awareness/self-awareness, purposiveness, emotional/responsiveness, imagination, conscience/ethics, and so on. Robots do not possess all these features together, for instance they may have some intelligence, some partial autonomy, or some degree of self-awareness, while other features, such as empathy and free will are very difficult (if not impossible) to be embedded into robots.

It has been claimed that robot soldiers can be designed to distinguish combatants from civilians more accurately than human soldiers, or that can behave more ethically than human soldiers. These claims have been challenged by many roboticists. The principal objection is that with current technologies we cannot create conscious robots. Therefore, in this case there would be moral production without moral consumption.

The benefits of pursuing research on robot conscience and ethics that could lead to the creation of an ethical robot, include the following:

- Working on robot ethics will require full clarification what it means to behave ethically, thus advancing the study of ethics.
- A clear objective basis of ethics will be required to be found, that makes ethics computable.
- Research on robot/machine ethics may lead to the discovery of new problems with current ethical theories, and, sometimes, to the development of better theories.
- Ethicists working with machine intelligence have a greater chance to achieve break thoughts in ethics, than theoretical ethicists working alone.

The purpose of the present chapter is to provide an account of robot conscience and ethics including a number of fundamental concepts and aspects. Specifically, the chapter:

- Discusses the ontological aspects of human conscience, and robot conscience/morality.
- Presents the general issues of robot ethics (roboethics).

- Outlines the two basic approaches to roboethics, namely top-down and bottom-up roboethics.
- Outlines the ethics of human-robot symbiosis along with the issues of robot rights and robot moral status.
- Presents the fundamental ethical aspects and problems of medical, assistive, socialized, and war robots (including the major objections to the use of war robots).

8.2 What Is Conscience?

The popular meaning of conscience has been "the soft inner voice that tells us when we have done wrong". This was a cause of confusion because conscience is not an audible voice, but speaks to us in several ways. The essence of conscience is that it speaks to us the rightness and wrongness of our actions, it provides judgments in our actions, advising us whether they attain or fall short of their goals and purpose [1–5]. The Greek equivalent of conscience is the word *"συνείδησις" (syneidesis)* which is composed by two words (συν = con = shared/with) and (είδησις = knowledge). Thus, the original meaning of the word "syneidesis" was shared knowledge, but soon came to mean "something done with knowledge on the part of the individual". Conscience cames from the Latin word *"conscientia"* which had both meanings: "shared knowledge" and "something done with one's knowledge". According to Schneider "conscience is a relation of the meaning of one's actions and the following moral responsibility for them".

Conscience of a person has no sense without consciousness, although consciousness may exist without conscience for some individuals (surely a bad thing). Conscience is very important when we are selecting our goals and actions that agree with our conscience and penalizes those that disagree. In other words, concerning past actions, conscience will judge that they have been in harmony with our total being in all our relations with society and with our own convictions and destiny. In this case conscience will approve the actions, and a sense of fulfillment will come over us. On the contrary, conscience condemns a sense of frustration and disharmony will come over us, which is the conviction of guilt and remorse.

From the above we see that conscience is the inner mechanism which regulates one's actions, and includes orders and suggestions, i.e., some of them are compulsory and must not be disobeyed (e.g., religious laws and commandments), while others are only suggestions that one tries to follow as best as he/she can, because sometimes they may be contradictory.

Emotions are closely related to conscience, since they play a significant role in the management of conscience by humans. They help us to select decisions that are good for the individual and for the societies we belong to. Remorse is an emotion of primary importance for conscience, since it incites us to follow our conscience and avoid any forbidden action that will lead us later to feel this unpleasant emotion. Man is the only one of all animals who can reflect of himself and orient his gaze upon his own actions and pass rational and conscious judgment upon them. The difficult point here is to find how one can determine what is right, and what is wrong, and how to respect conscience. This can be done through the moral norms (ethical rules) which are the result of one's reasoning about the purpose of his/her life and/or the result of education on the part of the family and society in which he/she grew up, and/or through the authority of faith, if he/she accepts relevance. Thus, every person has a composite of subjective and objective norms. The study of these morals is the subject of ethics and ethics theories as discussed in Section 2.6.

8.3 Robot Conscience and Morality

The conscience of robots is determined by the values designed into them [5, 6]. Whether robots will be moral agents with human-like consciousness and conscience is a subject upon which theorists and thinkers disagree. Discussions about when artificial beings might be held responsible for their actions, and whether they might deserve rights under legislation have furthered the understanding of moral agency and legal responsibility (see Section 8.4.5). Engineers have always been concerned about the safety of the systems they build. The development of autonomous machines and robots making decisions that can increasingly affect humans for good or bad has given rise to the new field of machine/robot conscience, morality and ethics. Designers of modern intelligent autonomous and conscious robots cannot always predict how they will act in new circumstances with new inputs, and so safety and conscience require that robots will become moral agents. Therefore, philosophers and roboticists are cooperating to create robots that can make moral decisions. Designing machines and robots that will not harm humans is both a practical and philosophical challenge, and in any case a very difficult task.

Conscience of humans and robots are very different. A robot is built such as it is very serious for it to fail to solve a problem. Human goals and robot goals are not the same, although conscience in both cases is related to the need to achieve crucial goals. Therefore, in describing the conscience of a robot we must include all the aspects that favor the goals and the actions defined as good by its designer and leader, and avoid to commit actions that he considers as wrong. Some of the goals concern the achievement of a task, while other goals are set for ensuring that the robot can work correctly and failure free. Like human conscience, there is no universal conscience for machines and robots; it depends on its goals and its physical realization.

Robot conscience and morality address a number of questions that include the following:

- Why robots with morality?
- Are robots the kinds of beings that can in principle make moral decisions? If so, why? If not, why not?
- Is moral decision making something that can be implemented in robots? If so, how?
- When humans want machines and robots making moral decisions? In what circumstances?
- Whose or what morality should be implemented?

The above questions have received great attention in recent years within the new field of "robot ethics" *(roboethics)* which is the branch of applied ethics that also includes the ethics (conscience) of the people who create and employ robots, and the ethics of how people treat robots. Roboethics will be further discussed in the next section.

8.4 Roboethics

8.4.1 General Issues

The term *roboethics* for "robot ethics" was coined by Gianmarco Veruggio [7]. Roboethics addresses the fundamental issues of the dual-use of robots (robots can be used or misused), the anthropomorphization of robots, the humanization of human-robot symbiosis (co-existence and cooperation), the reduction of the socio-technological gap, and the effect of robotics on the fair distribution of wealth and power.

Some more specific questions that have to be considered in the framework of roboethics are:

- What role would robots have into our future?
- How might human ethics be extended in order to be applicable to the combined human-robot actions?
- Who or what is responsible if a robot causes harm?

- Are there any types of robots that should not be designed? If so, why?
- Are there risks in creating emotional bonds of humans with robots?

The three principal positions of robotics scientists about roboethics are [7]:

- *Not interested in roboethics* (Advocators of this position argue that the action of robot designers is purely technical and does not have a moral or social responsibility in their work).
- *Interested in short-term robot ethical issues* (This is the attitude of those who consider the ethical performance in terms of *good* or *bad*, and adopt certain social and ethical values).
- *Interested in long-term robot ethical issues* (Roboticists having this attitude express their robotic ethical concern in terms of global, long-term aspects).

In general, the technological and computer advancements are constantly continuing to promote reliance on automation and robotics. Robots live with people, and so the ethical examination of robot creation and use makes sense both in short-term and long-term. Roboethics is a human centered ethics. Therefore, it must be compatible with the ethical and legal principles adopted by the international human rights organizations. Robotics on its own is a very sensitive field since robots are closer to humans than computers (or any other artificial beings) that may ever be created both morphologically and literally. This is because of their shape and form which reminds us of ourselves. Scientists should have in mind that robots may influence how societies develop in ways that could not be anticipated during their design process. Clearly, the more intelligence and autonomy is embedded into a robot (with the power to exert it) the more moral and ethical sensitivity is required [8]. Today there is no special legislation about robots (especially in the West).

Robots are legally treated in the same way as any other technological artifact and equipment. This is probably due to the fact that robots with full cognition, intelligence and autonomy are not yet in operation or in the market. However, many roboticists argue that even with the present pace of artificial intelligence and robotic technology, such laws will be required soon.

It is remarked that the primary requirement from a robot (and any other autonomous agent) is not doing harm. The issue of resolving the vague moral status of moral beings and human ethical dilemmas or ethical theories is a must but at a secondary level. This is because as robots get more complexity and capabilities it will become necessary to engineer more sophisticated safety control measures and mechanisms that can prevent the most critical dangers or potential harms.

Naturally, different cultures and societies have different views of autonomy, dignity and conscience. Thus, possible "ethical robots" of the future that may be developed in societies with different cultures would have embedded different ethical codes, something which complicates the transfer of robots from one culture to another (although their codes would have much in common). Over the years a large number of papers and books have been published dealing with particular robot conscience and ethics questions, adopting different moral views. Due to the fact that it is difficult to define universally acceptable ethical principles, it seems very difficult (if not impossible) to homogenize all these attempts.

As *Wendell Wallach* explains, the implementation of robot ethics can be done in two basic ways, namely [8]:

- *Top-down approach* (A certain number of ethical rules or principles governing moral behavior are prescribed and programmed in the robot system).
- *Bottom-up approach* (A developmental or evolutionary psychological-like way is followed to learn proper responses to moral considerations. This is analogous to how growing children learn conscience, i.e. what is right and what is wrong, based on social context and experience).

8.4.2 Top-Down Roboethics Approach

The top-down approach can be applied to both deontological and utilitarian/consequentialist theories.

Deontological robotethics In Section 2.6 we have seen that in deontological ethics the actions are evaluated for their own, rather than by the utility value or the consequences they produce. Actions implement "moral duties" which are determined by conscience (what is right, or wrong) independently of the actual consequences may cause.

In robotics the first ethical system proposed is that of the three Asimov's laws [9]:

Law 1: A robot may not injure a human being or, through inaction, allow a human being to come to harm.

Law 2: A robot must obey orders it receives from human beings, except when such orders conflict with the first law.

Law 3: A robot must protect its own existence as long as such protection does not conflict with the first or second laws.

Later, Asimov added a law which he named "Law Zero", since it has a higher importance than laws 1 through 3. This law states:

Law 0: No robot may harm humanity or, through inaction allow humanity to come to harm.

These laws are human-centered, i.e., they consider the role of robots in human service and imply that robots have sufficient intelligence and consciousness to make moral decisions following the rules in all situations how matter their complexity. But, at least for now, we couldn't program a machine with Asimov's laws if we tried. Furthermore, we are a long way from constructing a robot that can fully anticipate the consequences of its actions (or inactions). In other words, given the current maturity level of cognitive and intelligent robots these laws, despite their elegancy and simplicity, cannot at present provide a practical basis for roboethics [10].

In [11, 12] it is argued that for a robot to be ethically correct the following conditions (desiderata) must be satisfied:

D_1: Robots only take permissible actions.

D_2: All relevant actions that are obligatory for robots are actually performed by them, subject to ties and conflicts among available actions.

D_3: All permissible (or obligatory or forbidden) actions can be *proved* by the robot (and in some cases, by associated systems, e.g., oversight systems) to be permissible (or obligatory or forbidden), and all such proofs can be explained in ordinary English.

The above ethical system can be implemented in top-down fashions [12].

Consequentialist roboethics   In consequentialist theory, morality of an action is evaluated by its consequences. The best present moral action is the action that leads to the best future consequences. In utilitarian theory the best future consequences are determined or predicted using a certain "goodness or utility measure". Actions are moral if they are oriented towards promoting the best-long term interests (greatest good, pleasure, happiness) for everyone concerned.

For a robot to be able to reason and act along the consequentialist/utilitarian ethics, it must have the following abilities:

- The ability to describe every situation in the world.
- The ability to produce alternatives.
- The ability to predict the situation(s) that would be the consequence of taking an action given the present situation.
- The ability to evaluate a situation in terms of its goodness or utility measure.

These requirements do not necessarily mean that the robot must have high-level artificial intelligence features, but high-level computational ones. Actually, many evaluation criteria were proposed over the years the most common of which is a normalized weighted sum of the various aspects of goodness (pleasure, etc.) for a given person. In ideal (universalist) utilitarian approach the weights for all persons are equal (all persons count equally). In an egoist approach, the weight for the egoist person is 1, and for all other persons is zero. In an altruist approach, the weight for the altruist is zero and positive for all other persons. The common objection to utilitarianism is that it is not necessarily just, since the goodness of the society as a whole does not guarantee that fundamental human rights and goodness of each individual will be respected [13], and so it is of limited acceptability [14]. One way to address this problem is to count more (i.e., assign greater weights) to the less fortunate persons than the more fortunate ones.

Many statistical studies have revealed that only a few people conform to the universalism ideal (all weights equal). For instance, most people give greater weight tevalue to their relatives or to people they know better. The weight selection method depends on the agent's value or axiology beliefs.

8.4.3 Bottom-Up Roboethics Approach

In this approach the robots are equipped with computational and AI capabilities to adapt themselves in some way to different contexts, such that to be able to act properly in complicated situations. In other words, the robot becomes able to learn starting from perception of the world using a set of sensors, proceeds further to the planning of actions based on the sensory data, and then finally executes the action [15]. Very often, the robot is not going directly to the execution of the decided action, but via intermediate corrections. The process is similar to the way children learn their ethical performance from their parents through teaching, explanation, and reinforcement of good actions. In overall, this kind of moral learning falls within the framework of *"trial and error"*. In the literature there are many examples of robots created using this "developmental and tutelage process" (e.g., [16, 17]).

It was argued by many roboticists that to create an ethical robot both top-down and bottom-up approaches are required to be jointly applied (hybrid moral learning). Obviously, this hybrid approach is more powerful since the top-down principles are used as an overall guide while the system has the flexibility and moral adaptability of the bottom-up approach.

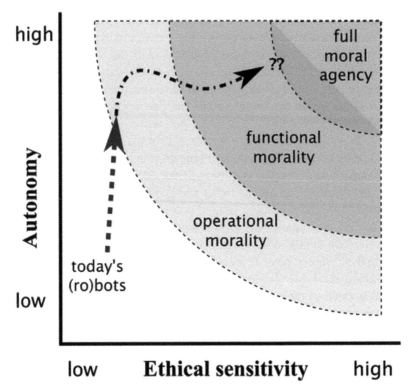

Figure 8.1 The three levels of morality vs. robot autonomy.

Source: http://cdn.wonderfulengineering.com/wp-content/uploads/2014/05/fulllength-robotics-1.png

In [8] it is argued that robot morality can be distinguished in three levels (Figure 8.1).

- *Operational morality* (The moral significance and responsibility lies entirely to the robot designers and users).
- *Functional morality* (The robot has the ability to make moral judgments when deciding a course of action without direct top-down instructions from humans. In this case the designers can no longer predict all the robot's actions and their consequences).
- *Full morality* (The robot is intelligent and entitled to select its own actions autonomously, and so it is fully responsible for them. At present no robot has been developed with full morality).

We again remark that actually robot moral decision making is a natural extension to engineering safety for systems with greater intelligence autonomy and consciousness. No doubt, as robots become faster, more intelligent, more autonomous, and more powerful, the need to endow them with a sense of conscience and morality becomes more and more urgent. A robot example claimed to have its own conscience is the Spanish android robot named **AISoy 1** (developed by AISoy Robotics Company). This robot has the ability to learn from experience and modify its behavior values and actions that can undertake at specific times [18]. If AISoy 1's are given to two different families, within two months they will be totally different because they will have had different experiences.

8.4.4 Human – Robot Symbiosis Roboethics

A very important issue in roboethics is connected to the human-robot symbiosis (living together, partnership) which aims to fill in the gap between fully autonomous and fully human-controlled robots. This is done by a proper dynamic subdivision of the human and the robot work, to optimize the admissible task range, accuracy, and work efficiency of the human-robot system (shared autonomy) as described in Chapter 6. From an ethics point of view, in addition to the fundamental issues for the shared autonomy (human-robot architecture, human-robot communication, autonomous machine learning, autonomous task planning and execution monitoring), the human agent must be supported by a number of specialized agents, which include the following [19]:

- *Interaction agent* which can handle more pro-active functions of the interaction (e.g., communication) and model the interaction with the help of the ethical/social agent.
- *Ethical/Social agent* which contains a set of social and ethical rules of interaction that enable the robot to execute ethical actions and interact with people according to the accepted ethical and social context.

Humans cooperating with robots must not be viewed as components of the system in the same way as the robots or the computers. Human-machine symbiosis is a subject of study for long time. For instance in [20] a vision for human-machine symbiosis is formulated as follows:

"Men will set the goals, formulate the hypotheses, determine the criteria, and perform the evaluations. Computing machines will do the routinizable work that must be done to prepare the way for insights and decisions in technical and scientific thinking".

Some philosophical questions about human-robot symbiosis are the following:

- What is the meaning of the master-slave human-robot relation?
- What is the meaning of robot as a partner in different contexts?
- How building social robots shapes our self-understanding, and how these robots impact on our society?

8.4.5 Robot Rights and Moral Status

Another issue related to roboethics is the issue of robot rights. Current Western legislation considers that robots are *objects* (inanimate agents) without duties or rights. Robots and computers are not legal entities and have no standing in the juridical system, and so they may not be perpetrators of a crime. A human that dies in the arms of a robot has not been murdered. But what should happen if a robot has partial consciousness and sense of self-awareness and self-preservation, and makes moral decisions? Should such moral robots possess the rights and duties of humans or some other rights and duties? This question has caused much discussion and debate among roboticists, sociologists, and philosophers.

There are many roboticists who envision that after some decades robots will be sentient and so they will need protection, arguing that a *"Bill of Rights for Robots"* must be developed. They argue that humans need to exercise empathy to the intelligent/conscious robots they create, and robots must be programmed with a sense for empathy for humans (their creators) [21]. The hard question here is: "How could we be able to recognize that the robot is truly conscious and moral, and there is not something deceptive about an ethical robot? Will humans think that such a robot has qualities that it most likely will not possess?"

In [22] it is argued that "the ethical aspect of an ethical robot does not necessarily entail a deception", since "for a human to accept that a robot is following ethical principles, it is only essential that it is able to justify its actions by appealing to acceptable ethical principles". This means that far from being deceptive the moral aspect of a robot should be transparent to its human partner such that its actions would not appear arbitrary.

An objection to this point of view might be that human will still need to believe that the robot has feelings or emotions, to accept it as a moral being, because only a being with own feelings could appreciate the feelings of others. For this point, in [23] it is maintained that "it is not clear that a robot could not be trained to take into account the suffering of others in calculating how it should behave in an ethical dilemma without having feeling itself". Feelings and emotions may sometimes *"drive away"* a human to the point where he is incapable to act morally, i.e. emotionality can be regarded as a "drawback" of humans that often prevents them to follow conscience. As stated in [23], once this is recognized, humans would not expect an ethical robot to be "emotional". On the contrary, humans may actually be glad that robots do not have feelings or emotions.

In [24] it is argued that "if a robot is built in future with humanlike capabilities that might include consciousness, there would be no reason to think that the robot has no real consciousness. This may be regarded as the starting point of assigning legal rights to robots".

In [25] an argument for the possibility of extending legal protection to robotic companions (social robots) is presented which is based on the view that the law is actually a *"social contract"*. Laws aim to be used to govern behavior for the greater good of society (i.e., laws must be used to influence people's preferences, rather than the opposite) in a utilitarian way. Thus, if the law's purpose is to reflect social norms and preferences, the desire of the society for robot rights (if it exists) should be taken into account and converted to law.

At the other end many roboticists and thinkers argue strongly against giving to robots moral or legal responsibility or legal rights. In [26], it is stated that robots are fully owned by us, and the potential of robotics should be interpreted as the potential to extend our abilities for achieving our goals. The interpretation of the statement "robots should be slaves" in no way means that "robots should be people you own", but "robots should be servants you own". In summary, the thesis presented in [26] is: "Robots are tools, like any other artifact when it comes in the domain of ethics. An autonomous robot definitely incorporates its own motivational structure and decision mechanisms, but *we* choose those motivations and design the decision making system. All their goals are derived from us.... We are not obliged to the robots, but to the society".

Closing our discussion on the issue of robot rights and moral status, we mention that Korea has developed a code of ethics for human-robot Interaction (the Robot Ethics Charta) which specifies ethical standards that would be programmed into robots, and limit some potential abuses of robots by humans [27]. Further issues on the morality status of artificial agents are discussed in [28, 29].

8.5 Medical Roboethics

Medical roboethics combines the principles of medical ethics and roboethics. The dominant branch of medical roboethics concerns the field of robotic surgery. The advocators of robotic surgery argue that robots assist surgeons to perform surgery more successfully, with enhanced access, visibility and precision. The basic advantages of robotic surgery include the reduction of pain and blood loss, reduction of hospital stays, and return of patients at home and normal life more quickly. However, there are many medical professionals that do not agree with the above benefits of robotics surgery. Another important area of medical (health care) robotics is the "rehabilitation and assistive robotics" which deals with the assistance, via robotics, of persons with reduced mobility and the elderly to improve their mobility and other physical capabilities. Assistive roboethics will be discussed in the next section.

8.5.1 Medical Ethics

Medical ethics (or health care ethics) is a branch of applied ethics that refers to the fields of medicine and health care. The origin of medical ethics goes back to ancient Greece and the work of Hippocrates who formulated the well known *"Hippocratic Oath"* [30]. Today there are modern versions of this Oath (ethical code) that use the term *maxim ("do no harm")*. The central issue of medical ethics is that medicine and health care deal with human health, life, and death. Over the years, ethicists, philosophers and health care givers (doctors, nurses, physical therapists) have attempted to provide principles for medical ethics. All classical applied ethics principles (utilitarian/teleologic, deontological, prima facie duties, and casuistry theory) find proper application (typically in combined way) in medical ethics.

Today medical ethics is practiced using the six-part *("George town Mentra")* ethics system which involves the following principles:

- *Autonomy* (The patient has the right to accept or refuse his/her treatment).
- *Beneficence* (The doctor should act in the best interest of the patient).
- *Non-maleficence* (The therapist should "first not to do harm").
- *Justice* (The distribution of scarce health resources and decision of who gets what treatment should be just).
- *Truthfulness* (The patient should not be lied and deserves to know the whole truth).
- *Dignity* (The patient has the right to dignity).

The above principles do not give by their own the answers as to how to resolve a particular situation, but may provide practical guide to doctors how they are obliged to morally treat real situations. Doctors should have in mind all the ethical theories and codes, trying to face any contradictory cases and dilemmas in the best moral way. For instance, an ethical dilemma occurs when the patience (or the person responsible for the patient's care) refuses life-saving treatments in which case a contradiction between autonomy and beneficence occurs.

8.5.2 Robotic Surgery Ethical Issues

Robotic surgery is a technique in which the surgeon performs surgery with the aid of a robot equipped with proper small dexterous tools. It can be used in several surgical procedures such as radical prostatectomy, mitral valve repair, coronary artery bypass, kidney transplant, hip replacement, hysterectomy, etc.) [31–33]. Figure 8.2 shows the world-wide surgery robot *DaVinci*.

Robotic surgery is not suitable for some complicated procedures, such as in certain kinds of heart surgery that require greater ability to move instruments in the patient's chest. Actually, robotic surgery covers the entire operating procedure from the acquisition of data to the surgery and post-operative examination. Typically, robot-assisted laparoscopic surgery allows a less-invasive procedure that before was only possible with more invasive open surgery. The selected operation plan is correlated with the intraoperative phase. The robotic system assists in guiding the movement of the surgeon to achieve precision in the planned procedure, and in many cases (e.g., hip replacement) the robot can work autonomously to perform part of or the entire operating procedure.

Robotic surgery ethics includes at minimum the principles and guidelines discussed in Section 8.5.1, and first of all should be legal, although a legal treatment may not always be ethical. The ethical standards are determined by the principles so far discussed, and, in the context of licenced doctors and surgeons, are provided by the accepted ethical code of medicine.

Discussions of a variety of surgical situations occurred over the years and their legal or ethical implications can be found in the medical literature [34, 35]. To take an idea of what should be happen in a robotic surgical operation we give here an example concerning the legal part of injuring a patient in a surgery of pangreatic tumor.

The law enforces on all individuals a duty of reasonable care to others, and determines it as law "a reasonable (rational) person in the same situation would act". If a person causes injury to another, because of unreasonable action, then the law imposes liability on the unreasonable person. In a surgery malpractice suit, a plaintiff would try to establish that the surgeon's actions were at odds with the standards accepted by the medical community, in order to prove that he breached his duty of care to the patient.

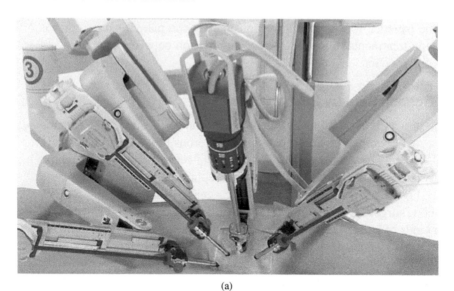

(a)

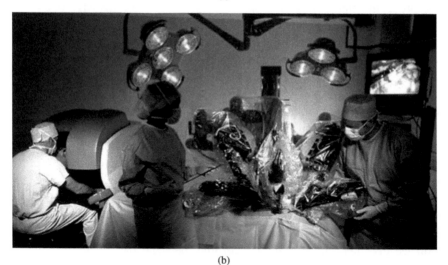

(b)

Figure 8.2   The Davinci surgical robot.

Source: (a) http://vrworld.com/Data/2010_9_23/GPGPU-Enables-Beating-Heart-Surgery/gpU_DaVinci_MIS_Robot_675.jpg

(b) http://bbsimg.ngfiles.com/1/25590000/ngbbs559c1d30a0f3a.jpg

The robotic surgery scenario which will be discussed in the following goes beyond the bounds of current personal injury law [36]. The surgeon explains to the patient the pangreatic tumor surgical procedure. The patient has provided informed consent for minimally invasive (laparoscopic) surgery, with the aid of a surgical robot (despite the risks involved in robotic surgery), and open surgery. The surgeon starts the surgery laparoscopically, and finds that the tumor cannot be removed by conventional laparascopic surgery. But on the basis of his past experience he believes that the robot with its greater dexterity can safely remove the tumor, which is actually the purpose of the robot. The surgeon setups and calibrates the robot and starts the operation of removing the tumor robotically when the robot malfunctions and injures the patient. The patient survives the operation but dies from cancer shortly after.

In case the patient's estate requests recovery of damages for the injuries due to surgery the following ethical issues arise [36]:

- *Has the surgeon acted ethically in offering the robotic surgery as an option for the patient knowing the inherent risks?*

To answer this question the medical code of ethics should be involved. In U.S.A. the *informed consent principle* is applied which states that "the physician's obligation is to present the medical facts accurately to the patient (or to the individual responsible for the patient's care) and to make recommendations for management in accordance to the medical practice. The physician has also an ethical obligation to help the patient make choices from among the therapeutic alternatives consistent with good medical practice". A surgeon is not obliged to ask the patient to indicate which surgical instrument prefers to be used, but under accepted standards in surgery it would not be ethical for the surgeon not to tell the patient about the use of the robot which differs very much from convention.

- *Has the surgeon acted ethically in deciding to use the robot?*

To answer this question (which does not differ from the same question in other medical malpractice cases) we have to look what a reasonable surgeon would have done in the same situation.

- *Who should be legally liable for the patient's injury?*

This case is complicated because of the death of the patient of his cancer. But, the patient would have die of cancer in the same way as he did. In fact he died after the robot was used.

If the patient's estate could sue for injuries incurred to the patient during the operation, the surgeon and the hospital would possibly seek indemnification from the robot manufacturer, sustaining that it was the robot's faculty behavior the cause of the injury. Then the manufacturer would likely sustain that the surgeon should not have opted to use the robot in this case and thus the surgeon assumed the risk of injuring the patient.

Other legal and ethical issues related to the above scenario would be:

- The responsibility of the manufacturer to assure that operators of the robot are adequately trained.
- The duty of the hospital to allow properly credential surgeons to use the robot.

8.6 Assistive and Socialized Roboethics

8.6.1 Assistive Robots

Assistive robots is a special class of medical service robots that are designed for people with special needs (PwSN) in order to assist them to enhance their mobility and attain their best physical and/or social functional level. People with loss of upper-limb control can function with the aid of manipulation robotic aids, and control an electric wheelchair using appropriate joysticks. Typical cases are tetraplegic patients due to spinal cord injury and quadriplegic due to other pathologies that lead to motor impairment of the four limbs. People of this category can benefit by the use of semi-autonomous mobility aids via proper human-machine interfaces [37].

The assistive robotic devices include the following:

- Assistive robots for people with impaired upper limbs and hands.
- Assistive robots for people with impaired lower limbs.
- Rehabilitation robots for upper or lower limb.
- Orthotic devices.
- Prosthetic devices (cyborg extensions).

Upper-limb assistive robotic devices are designed to assist persons with severe disabilities to perform everyday functions such as eating, drinking, washing, shaving etc. A modern service-assistive robot with 5-finger human-like hand is shown in Figure 8.3. Another robot for people with upper-limb disfunctions is the MANUS manipulator, which can also be mounted on a wheelchair [38, 39].

Upper-limb rehabilitation robotic devices are used for the evaluation and therapy of arms impaired as a consequence of stroke [40]. To get the best results a careful deep evaluation is needed in each particular case in order to use the most appropriate available device. In general, using therapeutic robots in the rehabilitation process, specific interactive and intensive training can be given.

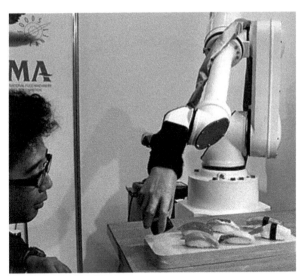

Figure 8.3 The Koii Sasahava/AP upper limb assistive robot.

Source: http://www.blogcdn.com/www.engadget.com/media/2009/06/6a00d8341c5d3253ef01156ff5a91
1970c-550wi.jpg

Lower-limb assistive robotic mobility devices include robotic wheelchairs and walkers. A robotic wheelchair of this kind with a mounted robotic manipulator is shown in Figure 8.4.

Some other intelligent/autonomous/semi-autonomous wheelchairs, are *SENARIO* (National Tech. Univ. of Athens/ZENON, Athens, EU Tide Project) [41], **VAHM** (University of Metz, France) [42], **ROLLAND** (Bremen Inst. of Safe Systems, Germany) [43], and **Maid** (FAW University of Ulm, Germany) [44].

Robotic walkers are designed for people who have some basic physical and mental abilities to perform a task but inefficiently and unsafely. A robotic walker can help a person to navigate and avoid collisions with obstacles, thus helping to reduce health costs and increase the quality of care and independency of handicapped people. A well-known robotic walker is the Cool Aide Robot Walker (Figure 8.5). Another walker is the "Veteran Affairs Personal Adaptive Mobility-Aid (VA-PAMAid) commercialized by Haptica Company [45].

Orthotic devices are used to assist or to support a weak and innefective muscle or limb. Typical orthotic devices take the form of an exoskeleton, i.e., a power anthropomorphic suit that is worn by the patient. Exoskeleton devices have links and joints corresponding to those of the human, and actuators.

Figure 8.4   A semi-autonomous wheelchair with mounted manipulator.

Source: http://3.bp.blogspot.com/-S6pwSEPkVl4/VEqulxqQv0I/AAAAAAAAANA/6a6Q4PpHeUo/
s1600/JACO%2BRobot%2Breach%2Bfrom%2Bchair.jpg

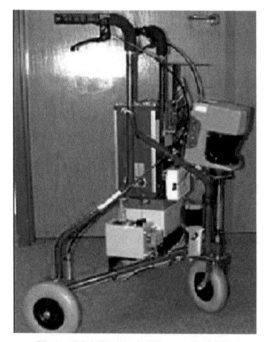

Figure 8.5   The Cool Aide robot walker.

Source: http://www.hizook.com/files/users/3/COOL_Aide_Robot_Walker.jpg

One of the early orthotic devices is the "wrist hand orthotic" **(WHO)** device which uses shape memory alloy actuators for providing a grasping function for quadriplegic persons [46]. Figure 8.6a shows an example of adjustable 'Wrist Driven Wrist Hand Orthosis' (WD-WHO) Device (North Coast Medical). Figure 8.6b shows a type of orthotic leg.

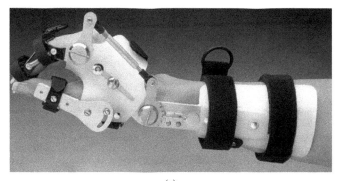

(a)

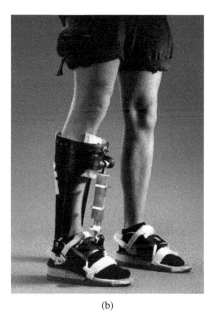

(b)

Figure 8.6   An orthotic exoskeleton leg.

Source: (a) https://ncmedical.com/item_1140.html

(b) http://gtalumnimag.com/wp-content/uploads/2012/08/office-space-feet-of-engineering-2_372x500.sh kl_.jpeg

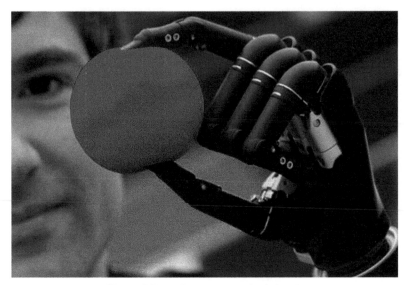

Figure 8.7   A dexterous prosthetic hand.

Source: http://static.ddmcdn.com/gif/5-artificial-limb-i-limb-660.jpg

Prosthetics are devices that are used as substitutes for missing parts of the human body. These devices are typically employed to provide mobility or manipulation when a limb is lost (hence the name artificial limbs). Figure 8.7 shows a prosthetic upper limb device.

Most people who wear prosthetic limbs are able to return to their previous activity levels and life styles (of course with hard work and determination). Several studies have revealed that there is a clear difference between elderly and younger or handicapped in accepting or rejecting assistive devices. The elderly tend in general to refuse technical aid, and prefer human help.

8.6.2 Socialized Robots

Socialized robots are intelligent (and conscious) robots used as social partners (for therapy or entertainment). The interaction of a human with a robot appears to be like interacting with another socially responsive being that cooperates with him/her as partner. Of course, it should be remarked that at present full human-like social interaction has not yet been achieved. In all cases there is some degree of shared control. According to the "American Food and Drug Administration, socialized robots are labeled as *Class2 medical devices* like powered/autonomous wheelchairs.

In the literature, several types of socialized robots are available, including the following [47, 48].

- *Socially evocative* (e.g., toy-like or pet-like entertainment robots).
- *Socially commutative* (These robots use human-like social cues and communication patterns that make the interactions more natural and familiar, and possess several levels of social-intelligence).
- *Socially responsible* (These robots can learn through human demonstration. They tend to be more perceptive of human social patterns passively, i.e., they respond to human interaction calls without been able to proactively engage with people in satisfying internal social aims).
- *Sociable* (These robots can proactively engage with people in order to satisfy internal social aims, e.g., emotions, drives, etc., including both the person's benefit and its own performance improvement).
- *Socially intelligent* (These robots, also called socially interactive robots possess several capabilities of human-like social intelligence using deep models of human cognition and social performance).

Socialized robots include also robots that can imitate humans. The two primary questions that have to be addressed for the design of such robots are [48]:

- How does a robot know *what* to imitate?
- How does a robot know *how* to imitate?

With reference to the first question the robot needs to detect the human demonstrator, observe his/her actions, and identify the ones that are relevant to the desired task. This requires the robot capability to perceive the human movement, determine what is important, and direct attention. With reference to the second question the robot needs, after the action perception, to convert this perception into a sequence of its own motor motions so as to get the same result. This is called the *correspondence problem* [49].

Examples of socialized robots are:

Kismet [50] An anthropomorphic robot head that has the ability to interact with humans physically, affectively and emotionally, and perform playful infant-like interactions that help children to achieve and develop social behavior.

Paro [51] A baby seal-like robot able to respond to its environment and humans interacting with it (e.g., petting, talking to it, etc.). Paro's capabilities are acquired via tactile sensors, light sensors, touch sensitive whilskers, sound and voice recognition and blinking eyes placed in a head that turns and seems to be tracking human motion and pay attention to a person interacting with it. Paro was created for use as companions of dementia and elderly people, responding positively to gentle touches and nice words that patients are drawn to it (www.paro.jp/english/movie.html)

Kaspar [52] An anthropomorphic socialized robot designed for inter-action with children suffering by autism (a disorder of neural development characterized by impaired social interaction and communication). Kaspar belongs to the class of socialized robots that are used for autism therapy following the "play therapy" concept, which helps in improving quality of life, learning skills, and social inclusion.

Cosmobot [53] A robot designed to help children with disabilities such as *cerebral palsy*. Several studies have demonstrated that interaction of cerebral palsy children with Cosmobot increased the strength of their quadriceps to a functional level within the statistical norm. The children are able to control the robot's head, arm and mouth movements, and to activate the wheels hidden under its feet to drive it forward, backward, left and right.

AIBO [54] This is a robotic dog capable of performing as companion and adjunct to therapy, especially for vulnerable persons. For instance, elderly with dementia improved their activity and social behavior with AIBO, and children (normal or autistic) have shown positive responses to AIBO in shared reading sessions. AIBO has moveable parts, and sensors for distance, acceleration, vibration, sound and pressure monitoring During interaction with children AIBO offers to them its paw, and can respond with pleasure (green light) or displeasure (red light) after some kinds of interactions.

Instantiations of Kismet, Paro, Kaspar, Cosmobot and AIBO are shown in Figures 8.8–8.10.

Figure 8.11 Shows the walking socialized humanoid robots QRIO and NAO.

Five other socialized humanoid robots, moving with wheels, namely: the University of Southern California robot **Bandit**, the **WAKAMARU** robot, the **Robovie** robot, the **Mobiserv** robot and the **Maggie** robot are shown in Figures 8.12 and 8.13.

The robots of Figures 8.8 through 8.13 were tested experimentally in various human-robot social interactions with children (normal and autistic) and dementia elderly. Examples are provided in [58] (Robovie), [59] (AIBO), [66] (KASPAR), and [61] (Paro).

(a)

(b)

Figure 8.8 The socialized robots Kismet (a), and Paro baby seal (b).

Source: (a) http://www.npowetenschap.nl/.imaging/stk/wetenschap/photo/media/wetenschap/noorder
licht/artikelen/2015/Juni/Robot-Kismet/original/Robot%20Kismet.jpg
(b) https://cafechottomatte.files.wordpress.com/2011/07/paro-wg2.jpg

(a)

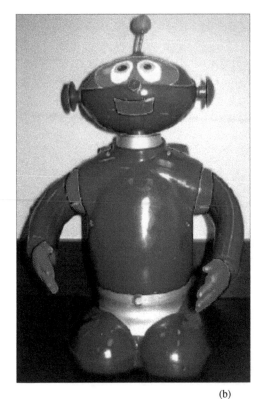

(b)

Figure 8.9 The socialized robots KASPAR (a), and Cosmobot (b).

Source: (a) http://www.roboticsbible.com/wp-content/uploads/2011/10/Robot-Kaspar.jpg
(b) http://socialrobots.org/images/cosmobot_image_desc.jpg

Figure 8.10 The robotic dog AIBO.

Source: http://www.about-robots.com/images/aibo-dog-robot.jpg

(a)

(b)

Figure 8.11 The walking humanoid robots QRIO (a), and NAO (b).

Source: (a) http://previewcf.turbosquid.com/Preview/2014/05/20_16_18_32/Sony_Qrio_Robot_Rigged_001.jpg900086e8-89ce-4340-be85-6829d1647833Large.jpg

(b) http://www.robotspodcast.com/podcast/mw_images/robotspodcast.com-Nao-Soccer-Robot2.jpg

http:// www.conscious-robots.com/images/stories/robots/aldebaran_nao.jpg

(a)

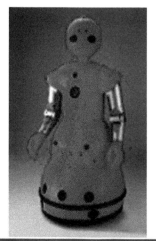

(b)

Figure 8.12   The wheeled humanoid robots Bandit (a), and Wakamaru (b).

(a)

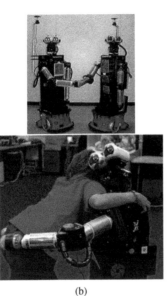

(b)

(c)

Figure 8.13 The wheeled humanoid robots Mobiserv (a), Robovie (b), and Maggie (c).

Source: (a) http://www.maxirobots.com/blog/wp-content/uploads/2013/12/mobiserv-06.jpg
http://soocurious.com/fr/wp-content/uploads/2013/08/ce-robot-domestique-verifiera-votre-etat-de-sante-et-pourra-meme-vous-sauver-la-vie-une.jpg
(b) http://www.irc.atr.jp/~kanda/image/robovie3.jpg
http://cdn.dogomedia.com/system/ckeditor_assets/pictures/511bb20d1860e0097b001391/content_atrgirl_robovie.jpg
(c) http://www.euron.org/images/robots/250_maggie2.jpg

8.7 Ethical Issues of Assistive and Socialized Robots

8.7.1 Assistive Robots

Assistive and socialized robotics are two classes of medical robotics, and so the principles outlined in Section 8.5.1 are applicable. Any assistive robot or device has the potential to be beneficially used or to be misused. The fundamental ethical questions surrounding the development and assistive robots are focused on human dignity, human relations, protection from physical/bodily harm, and the management health evaluation and other personal data. The Georgetown mantra guide principles are also valid, namely: autonomy, beneficence, non-maleficence, justice, truthfulness, and dignity. Other aspects that belong to the ethical obligation of the doctor/care giver include:

- Privacy
- Confidentiality
- Clinical accuracy
- Quality
- Reliability

All assistive technology programs should incorporate ethical statements into administrative policies and comply with professional codes of assistive ethics such as the "Rehabilitation Engineering Technology Society" ethical code **RESNA** [16] and the "Canadian Commission on Rehabilitation Councellor Certification" (**CRCC**) code [62, 63]. A general four-level ethical decision-making model for assistive and rehabilitation robotics is the following [64]:

Level 1: *Selection of proper assistive devices*
Employing inappropriate and counterproductive devices (and services) is a violation of the non-maleficence rules. The principles of beneficence justice and autonomy should also be respected at this level.

Level 2: *Competence of therapists*
This is the clinical multidisciplinary level, and is implemented via the effective cooperation of therapists. Some therapists may be more competent in using rehabilitation devices than other. The relevant Georgetown mantra rules should be adhered.

Level 3: *Efficiency and effectiveness of assistive devices*
Efficiency means the use of cost effective devices that are reliable and efficient. Here, the justice ethical rule is of the highest, priority, i.e. the PwSN persons must have their rehabilitation needs met. Professionals should be well educated in the up-to-date assistive technologies.

Level 4: *Societal resources and legislation*

This is the societal and public policy level which is operationalized by the legislative constituent relationship. Best practices rehabilitation interventions should be employed for all aspects.

The driving force should always be the quality of life improvement through the use of resources with maximum efficiency in conformity with the justice and autonomy principles.

8.7.2 Socialized Robots

These robots have to operate in actual environments that include family members, caregivers, and medical therapists. Typically, they do not apply any physical force on the user, although the user can touch it, often as part of the therapy. But in most cases no physical user-robot contact is involved, and often the robot is not even within the reach of the user. However, in the majority of cases the robot lies within the user's social interaction domain in which a one-to-one interaction occurs via speech, gesture, and body motion. Thus, the use of socialized robot raises a number of ethical concerns belonging to the psychological, social and emotional sphere. Of course, the Georgetown mantra medical ethics rules are also applied here as in the case of all assistive robots.

The two primary groups of socialized robots are children and elderly. Fundamental social and emotional (nonphysical) issues that need to be addressed when using socialized robots include the following [65]:

Attachment Emotional attachment can appear in all kinds of users (children, adult, elderly) and can create problems, e.g., when the robot is removed due to degradation or failures. The robot's absence may produce distress and/or loss of therapeutic benefits. This consequence can occur especially in users who cannot understand the reason of the robot removal (for instance, in dementia elderly).

Deception This risk can be particularly created when the robot is designed to physically mimic and perform like a human when acting in these roles. Robotic deception, occurring, for instance, when the patient perceives the robot as a doctor or nurse, may be harmful because the patient may believe that the robot can help him/her like a human (which is not true).

Awareness Both users and caregivers need to be accurately informed on the risks and hazards associated with the use of robots. The potential harm can be minimized by describing as much as possible the robot's capabilities and limitations to patients and caregivers as guidelines, possibly formalized as rules. Marketed robots are already covered by protection legislation.

Robot authority   A robot aimed to act as a therapist is given some authority to exert influence on the patient. Thus, the ethical question is who actually controls the type, the level, and the duration of interaction. For instance, if a patient wants to stop an exercise, due to stress or pain, a human therapist would reasonably accept this. Such a feature is desirable to be technically embedded to the robot, in order to balance ethically the patient's autonomy with the robot's authority.

Privacy   Securing privacy during human-robot interaction is of utmost importance. Patients expect to receive respect of their privacy, something which is typically backed by legislation. A robot may not be able to sufficiently discriminate information that should not be distributed (e.g., sensitive personal data), or distinguish between authorized persons to get patient's delicate information. Therefore, the patient has the ethical and legal right to be properly informed on the robot's abilities, including visual ones via cameras mounted on the robot and transmission of acquired images to other agents.

8.8 War Roboethics

8.8.1 War Ethics

According to the war philosopher Carl von Clausewitz "war is the continuation of policy by other means", i.e. war is about governance using violence and not peaceful means to resolve policy that regulates life in a territory. L.F.L. Oppenheim defines war as "…a contention between two or more States, through their armed forces and imposing such conditions of peace as the victor pleases" (quoted in British Manual of Military Law, Part II). The deep cause of war is undoubtedly due to the human drive for dominance over the others. Humans have been fighting each other since the prehistoric times, and people have been discussing its rights or wrongs for almost as long. War is a bad thing and controversial social effects raise critical moral questions for any thoughtful person. The answers to these questions are attempted by war ethics. The ethics of war aims at resolving what is right or wrong both for the individual and the states (countries), contributing to debates on public policy, and ultimately at leading to the establishment of codes of war [66, 67].

Fundamental questions that have to be addressed are:

- Is war always wrong?
- Are there situations when it might be justified?

- Is war a result of intrinsic human nature or rather of a changeable social activity?
- Will war be always part of human experience or there are actions that can be done to eliminate it?
- Is there a fair way to conduct war, or is it all hopeless barbarous killing?
- After the end of a war how should post-war reconstruction proceed and who should be in charge?

The three dominant doctrines (traditions) in the "ethics of war and peace" are [68]:

- Pacifism
- Just war theory
- Realism

Pacifism A pacifist rejects war in favor of peace, so pacifism is anti-warism". A pacifist believes that war is always wrong. A pacifist does not object violence which does not lead to human killing, but rejects war believing that there are no moral grounds that justify resorting to war. Pacifism war criticized because it refuses to take ferocious measures required to protect individual persons and their country. By not resisting to international aggression, with effecting means, it ends up rewarding aggression and fouling to protect people who need protection.

Just war theory This is the theory that specifies the conditions for judging if it is just to go to war, and the conditions for how the war should be conducted. The founders of just war theory are: Aristotle, Plato and Cicero, and in recent times just war theory was endorsed by such notables as Augustine, Aquinas, Grotius, and Suarez. Just war theory tries to synthesize in a balanced way the following:

- Killing people is seriously wrong
- States are obliged to defend their citizens and justice
- Protecting innocent human life and defending important moral values sometimes requires willingness to use violence and force.

Just war theory involves three parts known by their Latin names, namely:

Jus ad Bellum Political leaders who inaugurate wars are responsible for obeying jus ad bellum principles, otherwise they commit war crimes. The Jus as Bellum principles are the following:

- *Just cause* (A war must be for a just cause. Fear with respect to a neighboring power is not a sufficient cause. Just cause involves defending against attack, recapturing things taken, correct public evil).

- *Right intention* (The only intention allowed for a country to go to war is for the sake of just cause. It is not sufficient to have the right reason. Seeking power, land grab or 'revence' do not constitute a right intention. Ethnic hatred or genocide are ruled out).
- *Legitimate authority and declaration* (For a country to go to war the decision should be made by the proper state's authorities declared publicly to the state's citizens and to the enemy state(s)).
- *Last resort* (a state should go to war only if it has tried every sensible non-violent alternative first).
- *Proportionality* (The war must be in proportion. Forces should be restrained to the minimal amount for achieving their goal. Weapons of mass destruction are typically seen as being out of proportion to legal ends).
- *Chance of success* (A country may go to a war only if it can foresee that doing so will have measurable impact on the situation).

Jus in Bello   It refers to justice in war, i.e., to conducting a war in an ethical manner. Military officers, commanders and soldiers have to be responsible for violation of the 'jus in bello' international war law (International Humanitarian Law, Geneva Conventions, Hague Conventions, and related international protocols).

The fundamental principles of the humanitarian "jus in bello law" are the following:

- *Discrimination* (It is immoral to kill civilians, i.e., non-combatants. Soldiers must discriminate between unarmed civilians and those legitimate military, political, and industrial targets involved in basic rights-violating harm).
- *Proportionality* (All over the war's duration, soldiers are entitled to use only force proportional to the goal sought. Blind bombing is not ethical and is not allowed).
- *Benevolent treatment of prisoners of war* (Captive enemy soldiers are no longer engaged in harm, and so they are to be provided with benevolent, not malevolent, quarantine away from battle zones, and they should be exchanged for one's own prisoners of war (POW), after the end of the war).
- *Controlled weapons* (Soldiers are allowed to use controlled weapons and methods that are "not evil in themselves". Genocide, ethnic cleansing, and poisonous weapons are not allowed).

- *No retaliation* (The history has shown that retaliations do not work and actually lead to an escalation of death and an increasing destruction of war. Willing well is the best "revence").

Jus post-Bellum This refers to justice at war termination. It is intended to regulate the termination of wars and facilitate the return to peace. The return to peace is left to moral laws (since no global international just post bellum law exists). Some of this moral laws (non-exhaustive) are:

- *Proportionality* (Peace recovery should be reasonable measurable, and publicly declared).
- *Rights vindication* (The basic human rights to life and liberty, and state sovereignty should be morally addressed).
- *Distinction* (Leaders, soldiers, and civilians should be distinguished in the defeated country that is negotiating with).
- *Punishment* (Punishments for war crimes apply equally to all sides of the war. Punishments for jus war rules violation include financial restitution subject to distinction and proportionality, reformation of institutions in an aggressor state, etc.).

From the above it is clear that a war is a *Just war* if it is both justified and carried out in the right way. It is noted that some wars conducted for just causes have been rendered to unjust during the war because of the way they were fought.

Realism Realists advocate that war is an inevitable process taking place in the anarchical world system. Classical realists include Thucydides (in ancient Greece), Maciavelli (a Florentine political philosopher), and Hobber (an English philosopher). Modern realists include George Kennan, Reinhold Niebuhr, and Henry Kissinger. The doctrine of realism is mostly influential to political scientists, scholars, and practitioners of international affairs. According to realists war must be launched only for the country's self-interest, and once it has started, a state must do anything it can do to win. They believe that in war "anything goes", i.e., during wartime international war laws are not applicable. Actually, there are two distinct types of realism, namely:

- *Descriptive realism* (The states cannot behave morally in wartime either for reasons of motivation or for reasons of competitive struggle).
- *Prescriptive realism* (A prudent state is obliged to act amorally in the international scene, because if it is too moral, other states will quite probably exploit it and act more aggressively).

8.8.2 The Ethics of Robots in War

Although war robots with full autonomy are not yet operating in war fields the benefits and risks of the use of lethal machines for fighting in wars are of crucial concern. The ethical and legal rules of conducting wars using robotic weapons, in addition to conventional weapons, include at minimum all the rules of war discussed in the previous section. The 'jus ad bellum', 'jus in bello', and 'jus post bellum' principles should be respected, but the use of semi-autonomous/autonomous robots add new rules and require special considerations. The four fundamental issues of robots in war are the following:

- Firing decision
- Discrimination
- Responsibility
- Proportionality

Firing decision At present, the decision to use a robotic weapon to kill people still remains with the human operator, not only because of technical reasons but also for ensuring that human remains "in the loop" [66]. However, it is a fact that the separation margin between human firing and autonomous firing in the war field is constantly decreased. As stated in [13], even if all war robots were to be supervised by humans one may still be in doubt to what extend this is actually so. Moreover, as stated in [69], it is not always possible to avoid giving full autonomy to the robotic system. According to the US Department of Defense Combat aircrafts must be fully autonomous in order to operate efficiently [69]. This is because some situations may occur so quickly and need such fast information processing that we would entrust the robotic systems to make critical decisions. If human soldiers have to monitor the actions of each robot as they take place, this may restrict the effectiveness for which the robot was designed [70]. But, even if an autonomous robotic weapon is not illegal on account of its autonomy, the *just war law (Rules of Engagement)* requires that targeting should respect the principles of discrimination and proportionality.

Discrimination This is the ethical and legal issue that has received most attention in the use of robotic weapons. It is generally accepted that the ability to distinguish lawful from unlawful targets by robots, might vary enormously from one system to another. Present day robots are still far from having visual capabilities that may faithfully discriminate between lawful and unlawful targets even in close contact encounter [71]. In [72], it is argued that autonomous lethal robots should not be used until it is fully demonstrated that the systems can precisely distinguish between a soldier and a civilian in all situations. But in [13], exactly the opposite id advocated, i.e., "although autonomous (unmanned) robotic weapons may sometimes make mistakes, in overall they act more ethically than human soldiers. In [13], it is also argued that human soldiers (even if ethically trained) have more tendency to perform wrongly in war, and find difficulty to face justly war situations. In [70], the thesis of [13] is accepted, because, as argued, human soldiers may perform irrationally when in fear or stress, which cannot occur in combat robots. Therefore "combat robots may act more ethically than human soldiers independently of the circumstances, and so there is a compelling reason to pursue their development and study their capacity to act ethically". In [73], it is argued that: "In any case, an autonomous robotic system might be inadequate and unlawful in its ability to distinguish civilians from combatants in operational conditions of infantry urban warfare, but lawful in battlefield environments with few, if any, civilians present. At present, no one seriously expects remotely-controlled or autonomous systems to completely replace humans on the battlefield. Many military missions will always require humans on the ground, even if in some contexts they will operate alongside and in conjunction with increasingly automated, sometimes autonomous, systems".

Responsibility   In all cases of using robots for industrial, medical, and service tasks, the responsibility assignment in case of failure is unclear, and needs to consider both ethical and legislation issues. These issues are much more critical in the case of war/lethal robots, which are designed to kill humans with a view to save other humans, whereas in medical robotics the robot is designed to save human lives without taking other lives. The responsibility question is to whom blame and punishment should be assigned for improper fight and unauthorized harms caused (intentionally or non-intentionally) by an autonomous robot [72, 74, 75]; to the designers, robot manufacturer, procurement officer, robot controller/supervisor, military commander, a state's president/prime minister, or to the robot itself? Perhaps a chain of responsibility would be a simple solution, in which case the commanding officer is finally responsible. However, the situation is complicated and needs to be discussed more deeply when the robot was given higher degree of autonomy, which makes it a partially or fully 'more being' in future. Two problems that may be encountered in wars using robots are [70]:

- *Refusing an order* If a robot refuses a command to attack a house because its sensors "see through the walls" that are many children inside and it was programmed to follow the rules of engagement to minimize civilian casualties, we ought to defer to the robot which may have more situational awareness or the commander who has issued a lawful command on the basis of the information he has? On the other hand if a robot refuses an order and produces more harm who is responsible in this case? If we give a robot the authority to refuse orders, can this may be extended to human soldiers violating the basic military principle to "obey orders?".

- *Consent by soldiers to risk* In many modern war cases a malfunctioning of a semiautonomous or autonomous robotic weapon has killed "friendly soldiers". Thus should soldiers be informed on the risks incurred when using autonomous weapons or working with dangerous items such as explosives? Does consent to risk has any meaning, if soldiers generally have not the right to refuse a work or war order?

Proportionality   This rule requires that even if a weapon meets the test of distinction, any weapon must also involve evaluation that sets the anticipated military advantage to be gained against the predicted civilian harm (persons or objects). The proportionality principle requires that the collateral harm to civilians must not be excessive relative to the expected military gain.

8.8.3 Objections about Autonomous War Robots

The use of robotic weapons is subject to a number of objections. The major three of them are:

- *Inability to program war laws* (Quite possibly fully autonomous weapons will never achieve the ability to meet the war ethical and legal standards. Artificial intelligence has overpromised, since no machine will be able through its programming to replace the key elements of human emotion. Thus, adequate protection of civilians in armed conflict can be ensured only if humans oversights robotic weapons).
- *Human out of the firing loop* (It is wrong *per se* to take human out of the firing loop. A machine, no matter how intelligent is, cannot completely replace the presence of a human agent who possesses conscience and the faculty of moral judgment. Therefore, the application of lethal violence should in no circumstances ever to be delegated completely to a machine).
- *Lowering the barriers of war* (The long-run development of autonomous robotic weapons, which remove human soldiers from the risk, diminishes the disincentive to resort to war. Politicians who feel to have a moral duty to protect lives of their soldiers may favor efforts to replace human

fighters with robots. Push-button or risk-free wars that result in racked metal instead of casualties – at least to the country that uses robots – may lessen the emotional impact that wars have currently on the people of that country).

Further concerns about robotic weapons are discussed in [76, 77]. A recent open letter calling for a *"ban on offensive autonomous weapons"* was presented at the International Joint Conference on Artificial Intelligence (Buenos Aires, Argentina, 28 July 2015) [78]. A large number of people signed it including more than 1,000 experts and leading roboticists signed it. In [79, 80] two opposite replies were published, namely:

- We should not ban "Killer Robots", and here's why (Evan Ackerman, 29 July 2015) [79].
- Why we really should ban autonomous weapons: A response (Stuart Russel, Max Tegmark and Toby Walsh, 3 August 2015) [80].
- *Ackerman* states that "I'm not in favor of robots killing people. If this letter was about that, I'd totally sign it. But that's not what it's about; it's about the potential value of armed autonomous robots, and I believe that this is something that we need to have a reasoned discussion about rather than banning" [. . .]. What we really need is a way of making autonomous armed robots ethical, because we're not going to be able to prevent them from existing".
- *Russell, Tegmark and Walsh* state among others: "Autonomous weapons are completely different from human soldiers and would be used in completely different ways [. . .]. The key issue is the likely consequences of an arms race – for example, the availability on the black market of mass quantities of low-cost, anti-personnel micro-robots that can be deployed by one person to anonymously kill thousands or millions of people who meet the user's targeting criteria".

Examples of robotic weapons are shown in Figures 8.14–8.16, namely a predator unmanned drone aircraft, a stealth unmanned drone helicopter, and a land battle autonomous robot.

8.9 Concluding Remarks

Conscience is our inner process that tells us whether what we are doing is right or wrong. When we are choosing our goals and the necessary actions that enable us to achieve these goals, conscience first favors the goals that agree with our conscience, and penalizes those that disagree, and then favors our conscious actions that are compatible and permitted by our conscience.

Figure 8.14   A predator unmanned aircraft drone.

Source: http://commanderinstruments.com/images/300_uav_predator.jpg

Figure 8.15   A stealth unmanned helicopter drone.

Source: http://www.wired.com/images_blogs/dangerroom/2011/06/090508-N-2821G-158-660x396.jpg

Figure 8.16   A modular advanced armed robotic system (MAARS).

Source: http://i.ytimg.com/vi/j0TMfavtg_Y/0.jpg

Consciousness and conscience belong to the mechanisms that create the behavior of humans. The same is also true for robots and other artificial beings. In this chapter we provided an overview of robot conscience and ethics, namely: (i) the ontological issues of human and robot conscience, (ii) the general issues of roboethics (focusing on the top-down and bottom-up approaches), (iii) the ethics of human-robot symbiosis along with the issues of robot rights and robot moral status, and (iv) the ethical aspects of medical, assistive, socialized and war roboethics.

Questions about conscience that need to be addressed more deeply include the following:

- Will mental robots need their own conscience and moral agency?
- Human beings often do not respect or modify the principles of their conscience. Will robots and other artificial beings do the same?
- Is it possible to build a high security control of a mental robot, such that will never modify a part of its memory and conscience?
- Instead of wondering whether mental robots must have of conscience, is it ethical to modify our conscience to take into account the existence of mental robot? For example may we treat them as objects?
- Can mental robots (personal robots, androids, etc.) have rights like us, or at least like animals?

References

[1] A.J. Bahm, Theories of conscience, Ethics, 75, 1965, 128–131.

[2] R.P. George, Conscience and its Enemies, Wilmington, DE, U.S.A.: ISI Books, 2013.

[3] L. Stout, Cultivating Conscience: How Good Laws Make Good People, Princeton, NJ, U.S.A.: Princeton University Press, 2010.

[4] J. Pitrat, Speaking about and acting upon oneself, LAFORIA Report 95/29, 1995. http://webia.lip6.fr/~pitrat/Speaking_about_oneself.doc

[5] J. Pitrat, Artificial Beings: The Conscience of a Conscious Machine, Hoboken, NJ., U.S.A., J. Wiley & Sons, 2009.

[6] C.H. Ramey, Conscience as a design benchmark for social robots, Proceedings of 15th IEEE Intern. Symposium on Robots and Human Interactive Communication (ROMAN 2006), Hatfield, U.S.A., 6–8 September, 2006, 486–491.

[7] G. Verrugio and F. Operto, Roboethics: A bottom-up interdisciplinary discourse in the field of applied ethics in robotics, Intern. Review of Information Ethics, 6(12), 2006, 2–8.

[8] W. Wallach and C. Allen, Moral Machines: Teaching Robots Right from Wrong, Oxford, U.K.: Oxford University Press, 2009.

[9] I. Asimov, "Runaround", Astounding Science Fiction (March 1942). Republished in: Robot Visions by Isaac Asimov, New York, NY, U.S.A.: Penguin, 1991.

[10] S.L. Anderson, Asimov's three laws of robotics, and machine metaethics, Machine Ethics, AAAI Fall Symposium. Technical Report FS-05–06, AAAI Press, 2005, 8–16.

[11] S. Bringsjord, Ethical robots: The future can head us, AI and Society, 22(4), 2008, 539–550.

[12] S. Bringsjord and J. Taylor, The divine command approach to robotic ethics, In: P. Lin, K. Abney and G.A. Bekey (eds.), Robot Ethics: The Ethical and Social Implications of Robotics, Cambridge, MA, U.S.A.: The MIT Press, 2012, 85–108.

[13] R. Arkin, Governing Lethal Behavior in Autonomous Robots, New York, NY, U.S.A.: Chapman and Hall, 2009.

[14] P. Quinn, Divine Commands and Moral Requirements, New York, NY, U.S.A.: Oxford University, Press, 1978.

[15] M. Dekker, Can humans be replaced by autonomous robots? Ethical reflections in the framework of an interdisciplinary technology assessment, Proceedings IEEE International Conference on Robotics and Automation, Rome, Italy, 10–14 April, 2000.

[16] A.L. Thamaz, Social learning mechanisms for robots, Proceedings 14th Symposium on Robotics Research (ISRR), Lucerne, Switzerland, Aug. 31–Sept. 3, 2009.

[17] A. Lockerd and C. Breazal, Tutelage and socially guided robot learning, Proceedings of IEEE/RSJ Intern. Symposium on Intelligent Robots and Systems (IROS-2004), Taipei, Taiwan, Oct. 18–22, 2004, 3475–3480.

[18] The World's first robot with a conscience www.unexplained-mysteries. com/forum/index.php? https://www.aisoy.com, www.archive.today/ oVUo

[19] K. Kawamura, T.E. Rogers, K. Hambuchen and D. Erol, Toward a human-robot symbiotic system, Robotics and Computer Integrated Manufacturing, 19, 2003, 555–565.

[20] J.R.C. Licklider, Man-computer symbiosis, IRE Transactions on Human Factors and Electronics, 1, 1960, 4–11.

[21] K.J. Moore, A dream of robot's rights. http://hplusmagazine.com/2011/ 08/29/a-dream-of-robots-rights

[22] S.L. Anderson and M. Anderson, Special Session on "Machine Ethics", American Philosophical Association 2006 Eastern Division Meeting, Washington, DC, December, 2006.

[23] S.L. Anderson and M. Anderson, The consequences for human beings of creating ethical robots, In: T. Metzler (ed.), Human Implications of Human Robot Interaction, New York, U.S.A., AAAI, 2007.

[24] P. Singer and A. Sagan, Do humanoid robots deserve to have rights? The Japan Times, 17 December, 2009. www.japantimes.co.jp/opinion/2009/ 12/17/commentary/do-humanoid-robots-deserve-to-have-rights

[25] K. Darling, Extending legal rights to social robots. http://ssrn.com/abstr act=2044797

[26] J.J. Bryson, Robots should be slaves, In: Y. Wilks (ed.), Close Engagements with Artificial Companions: Key Social, Psychological, Ethical, and Design Issues, Amsterdam, The Netherlands: John Benjamins, 2008.

[27] S. Lovgren, Robot code of ethics to prevent android abuse; protect humans, National Geographic News, May, 2007.

[28] L. Floridi and J.W. Sanders, On the morality of artificial agents, Minds and Machines, 14, 2004, 349–379.

[29] S. Torrance, Ethics and consciousness in artificial agents, Artificial Intelligence Society, 22, 2007, 495–521.

[30] M. North, The Hippocratic Oath (Translation), National Library of Medicine 2012. Greek Medicine: www.nlm.nih.gov/hmd/greek/gree_ oath.html

[31] B.H. Taylor, Computer-Integrated Surgery, Cambridge, MA, U.S.A.: MIT Press, 1996.

[32] R.M. Satava, Laparoscopic surgery, robots, and surgical simulation: Moral and ethical issues, Seminar Laparoscopic Surgery, 9(4), 2002, 230–238.

[33] G. Gromez, Emerging technology in surgery: Informatics, electronics, and robotics, In: C.M. Towensend, R.D. Beauchamp and B.M. Evers, Sabiston Textbook of Surgery, Philadelphia, PA, U.S.A.: Saunders Elsevier, 2007.

[34] A. Mavroforou, E. Michalodimitrakis, Legal and ethical issues in robotic surgery, International Angiology, 29, 2010, 75–79.

[35] L. Rogozea, F. Leasu, A. Rapanovicki and M. Baritz, Ethics, robotics and medicine development, Proceedings of 9th WSEAS Intern. Conf. on Signal Processing, Robotics and Automation (ISPRA'10), University of Cambridge, Cambridge, England, 2010.

[36] D.S. Kemp, Autonomous cars and surgical robots: A discussion of ethical and legal responsibility, Verdict: Legal Analysis and Commentary from Justia, 2012. http://verdict.justia.com/2012/11/19/autonomous-cars-and-surgical-robots

[37] N. Katevas, Mobile Robotics in Health Care (Ch.1), Amsterdam, The Netherlands: IOS Press, 2001.

[38] MANUS Robot: www.exactdynamics.nl

[39] http://ots.fh-brandenburg-de/downloads/scripte/ais/IFA-Serviceroboter-DB.pdf

[40] P. Maciejasz, J. Eshweiler, K. Gerlah-Hahn, A. Jensen-Troy and S. Leonhardt, A survey on robotic devices for upper limb rehabilitation, J. Neuro Engineering and Rehabilitation, 11(3), 2014, 1–29.

[41] N. Katevas, N.M. Sgouros, S.G. Tzafestas, G. Papakonstantinou, et al., The autonomous mobile robot SENARIO: A sensor-aided intelligent navigation system for powered wheelchairs, IEEE Robotics and Automation Magazine, 4(4), 1997, 60–70.

[42] G. Bourhis, O. Horn, O. Habert and A. Pruski, An autonomous vehicle for people with motor disabilities, IEEE Robotics and Automation Magazine, 7(1), 2001, 20–28.

[43] A. Lankenau and T. Röfer, A versatile and safe mobility assistant, ibid, 2001, 29–37.

[44] E. Prassler, J. Scholz and P. Fiorini, A robotic wheelchair for crowded public environments, ibid, 2001, 38–45.

[45] P. Lun, D. Reinkensmeyer, R. Mahoney, W. Rymer and C. Burgar, Robotic Devices for Movement Therapy after stroke: Current status and challenges to clinical acceptance, Top Stroke Rehabilitation, Winter, 2002, 40–53.

[46] About Your New WHO, Surgi-Care Inc. http://www. surgi-careinc.com/pi/about-your-new-who-wrist-hand-orthosis

[47] C. Breazeal, Towards sociable robots, Robotics and Autonomous Systems, 42, 2003, 167–175.

[48] K. Dautenhahn, Socially intelligent robots: Dimensions of human robot interaction, Philos.Trans.Royal Soc. London: B-Biological Sciences, 362(1480), 2007.

[49] C.L. Nehaniv and K. Dautenhahn, The correspondence problem, In: K. Dautenhahn and C.L. Nehaniv (eds.), Imitation in Animals and Artifacts, Cambridge, MA , U.S.A.: MIT Press, 2002, 41–61.

[50] C. Breazeal, Designing Sociable Robots, Cambridge, MA, U.S.A.: MIT Press, 2002.

[51] T. Shibata, T. Mitsui, T. Walda and A. Touda, Mental commit robot and its application to therapy of children, Proceedings IEEE/ASME Intern. Conf. on Advanced Intelligent Mechatronics (AIM'01), Como, Italy, 2001, 1053–1058.

[52] K. Dautenhahn, C.L. Nahaniv, M.L. Walters, B. Robins, H. Kose-Bagri, N.A. Mirza and M. Blow, KASPAR: A minimally expressive humanoid robot for human-robot interaction research, Applied Bionics and Biomechanics, 6(1), 2009, 1–29.

[53] A.J. Brisben, C.S. Safos, A.D. Lockerd, J.M. Vice and C.E. Latham, The CosmoBot system: Evaluating its usability in therapy sessions with children diagnosed with cerebral palsy, 2005, Also: Special Issue: "Robots", 12, April 14–16, 2005.

[54] A. Miklosi and M. Gácsi, On the utilitarian of social animals as a model for social robotics, Frontiers in Psychology, 3(75), 2012.

[55] T. Ishida, Development of a small biped entertainment robot QRIO, Micro-Nanomechatronics and Human Science, 2007, 2483–2488.

[56] R. Gockley and M.J. Mataric, Encouraging physical therapy compliance with a hands-off mobile robot, Proceedings 1st Intern. Conf. on Human-Robot Interaction (HRI-06), Salt Lake City, Utah, U.S.A., 2006, 150–155.

[57] R. Mead, E. Wade, P. Johnson, A. St. Clair, S. Chen and M.J. Mataric, An architecture for rehabilitation task practice in socially assistive human-robot interaction, Proceedings 19th IEEE Intern. Symposium on

Robot and Human Interactive Communication, Viareggio, Italy, 2010, 404–409.

[58] T. Kanda, H. Ishiguro, T. Ono, M. Imai and R. Nakatsu, Development and evaluation of an interactive humanoid robot: "Robovie", Proceedings IEEE Intern. Conf. on Robotics and Automation, St. Paul, MN, USA, 14–18 May, 2002, 1848–1853.

[59] G.F. Melson, P.H. Kahn, Jr., A. Beck and B. Friedman, Robotic pets in human lives: implications for the human-animal bond and for human relationships with personified technologies, J. Social Issues, 65(3), 2009, 545–567.

[60] B. Robins, K. Dautenhahn and P. Dickerson, From isolation to communication: A case study evaluation of robot assisted play for children with autism with a minimally expressive humanoid robot, Proceedings 2nd Intern. Conf. on Advances in Computer – Human Interactions (ACHI'09), Cancum, Mexico, Feb. 1–7, 2009.

[61] AIST: National Institute of Advanced Industrial Science and Technology (AIST), Paro found to improve brain function in patients with cognition disorders, Transactions of the AIST, September 16, 2005.

[62] RESNA Code of Ethics. http://resna.org/certification/RESNA_Code_of_Ethics.pdf

[63] CRC Certification www.crccertification.com/pages/crc-ccrc_code_of_ethics/10.php

[64] V. Tarvydas and R. Cottone, Ethical response to legislative, organizational and economic dynamics: A four-level model of ethical practice, J. Applied Rehab. Counseling, 22(4), 1991, 11–18.

[65] D. Feil-Seifer and M.J. Mataric, Ethical principles for socially assistive robotics, IEEE Robotics and Automation Magazine, 18(1), 2011, 24–31.

[66] P. Asaro, How just could be a robot war be? Amsterdam, The Netherlands: IOS Press, 2008, 50–64.

[67] BBC, Ethics of war www.bbc.co.uk/ethics/war/overview/introduction.shtm

[68] J.A. Coates, The Ethics of War, Manchester, U.K.: University of Manchester Press, 1997.

[69] www.defence.gov/Transcripts/Transcripts.apsx?TranscriptID=1108

[70] P. Lin, G. Bekey and K. Abney, Robots in war, In: R. Capuro and M. Nagenborg (eds.), Ethics and Robotics, Heidelberg: AKA Verlag Heidelberg, 2009, 49–67.

[71] N. Sharkey, The ethical frontiers of robotics, Science, 322, 2008, 1800–1801.

[72] N. Sharkey, Cassandra or false prophet of doom: AI robots and war, IEEE Intelligent Systems, 23(24), 2008, 14–17.

[73] K. Anderson and M. Waxman, Law and ethics for autonomous weapon systems: Why a ban won't work and how the laws of war can, Laws and Ethics of Autonomous Weapon Systems, Hoover Institution, Stanford University www.hoover.org/taskforces/national-security

[74] R. Sparrow, Killer robots, J. Applied Philosophy, 24(1), 2007, 62–77.

[75] P. Asaro, Robots and responsibility from a legal perspective, Proceedings of 2007 IEEE Intern. Conf. on Robotics and Automation workshop on Roboethics, Rome, Italy, 2007.

[76] P. Lin, Robots, Ethics and War, The Center for Internet and Society, Stanford Law School, November, 2012.

[77] F. Gros, C. Tessier and T. Pichevin, Ethics and authority sharing for autonomous armed robots, Proceedings of the 1st Workshop on Rights and Duties of Autonomous Agent (RDA2), Part of 20th European Conf. on Artificial Intelligence (ECAI 2012), Montpellier, France, Aug. 28, 2012, 7–12.

[78] http://futureoflife.org/AI/open_letter_autonomous_weapons#

[79] E. Ackerman, We should not ban "killer robots", and is why here. http://spectrum.ieee.org/automation/robotics/artificial-intelligence/we-should-not-ban-killer-robots

[80] S. Russell, M. Tegmark and T. Walsh, Why we really should ban autonomous weapons. http://spectrum.ieee.org/automaton/robotics/artificial-intelligence/why-we-really-should-ban-autonomous-weapons/?utm_source=roboticsnews&utm_medium=email&utm_campaign=080415

9

Architectural Aspects of Cognitive Robots

*A design isn't finished unless
somebody is using it.*
BrendaLaurel

*The more technology becomes complicated
inside the more it has to be simple outside.*
Derrick de Kerckhove

*The design process, at its best,
integrates the aspirations of art,
science, and culture.*
Jeff Smith

9.1 Introduction

Cognitive and intelligent control architectures constitute the *"core"* of modern mental robotic systems. The proper choice of architecture facilitates design, implementation and validation of robotic systems. The architecture concept involves various other concepts such as architectural structure and architectural style. The architectural structure has to do with the subdivision of a system in subsystems and their interconnection and interaction. The architecture style refers to the computational issues that are behind a given system (e.g., communication style, synchronous or asynchronous computation style, etc.). All cognitive and intelligent robotic systems embody some specific architectural structure and style or several structures and styles in a properly integrated form. The architecture and its implementation are usually very strongly tied together in a *"built it, and make it work"* way. Actually, different robotic applications have different needs which can be satisfied by a proper different architecture. However, selecting the proper architecture for a specific application is more an art than science, and hence only guidelines for this can be provided.

The principal steps for the design and implementation of any cognitive architecture are the following:

- Task definition.
- Task decomposition (in simple primitive/generic tasks).
- Task solving components' specification, development and implementation.
- Component development and implementation integration.
- Overall system integration, evaluation, and validation.

Actually, cognitive/intelligent robot architectures should include aspects that are constant over time and across different applications of the robot agent. Typically these aspects include the following:

- Short-term and long-term memories.
- Representation of the entities stored in the memories.
- Operations that are applied on these entities (including performance evaluation mechanisms and learning mechanisms that update them).

The purpose of this chapter is to provide an overview of principal architectural aspects of cognitive and intelligent robots. The chapter is technical and aims to serve as a tutorial introduction for those readers who are not merely interested in the philosophical aspects of mental robotics but also in their technical aspects. Specifically, the chapter:

- Provides an account of generic requirements and capabilities of cognitive robot architectures.
- Discusses two popular taxonomies of cognitive-intelligent robot architectures.
- Outlines some variants of meta-cognition in robot system architectures.
- Discusses the main conceptual and structural features of the MIT *Cog/Kismet,* and the *LAAS* hierarchical behavior-based architectures.
- Presents the *SOAR* and *ICARUS* cognivist (symbolic) architectures, the *Global Workspace* emergent architecture, and the *LIDA* hybrid (cognivist/emergent) architecture.
- Provides a list of 10 further cognitive-intelligent architectures with brief description and their references.
- Outlines the evaluation and validation architectural issues through several criteria, and exhibits the results of a comparative review of 20 cognitive architectures.

9.2 Features and Capabilities of Cognitive Robots Architectures

A preliminary discussion of features and design requirements of cognitive robots was given in Chapter 4 (Section 4.7). Here, we will provide a little more detailed account of an additional set of generic architectural requirements and capabilities. Central philosophical questions about cognitive architectures include the following:

- What *a priori* information does a cognitive robot need vs. the information that it can acquire during its operation?
- What knowledge does a cognitive robot need vs. what should be known by the designer?
- Under what conditions a robot should find out if something is valid by using perception as contrasted to reasoning based on its past experience?

The answers to these questions depend on the application where the robot is to be used as well as on the sophistication level adopted for the architecture.

The fundamental features that a robot cognitive architecture should have are [1]:

Knowledge representation   A robot encodes its knowledge using either a single uniform representation formalism or a combination of several formalisms. Using a single representation formalism provides simplicity and elegance and is more suitable for learning and reflection. Employing a mixture of formalisms provides some advantages but also, possibly, some drawbacks. Thus the selection of mixed formalism is crucial for the successful performance of the robot. The knowledge itself is not embedded into the architecture since it is not fixed and may change during the robot operation or across the application domains. Due to the change of knowledge over time, a good knowledge representation language must at minimum provide *fluents*, i.e., predicate or function symbols able to change their values according to the changes in the world. Examples of such representation languages include the situation calculus of *Ray Reiter* [2], modal vs. history-based and time-based vs. action-based languages, etc.

Knowledge organization   This refers to the way the knowledge is organized in the memory of the architecture. Typical forms of knowledge organization are the hierarchical and horizontal (flat) organizations. Examples of flat organizations are the production system formalism and the propositional logic formalism. Of course, these formalisms may influence one another. For instance, one production rule may lead to a different choice on the next cycle not directly, but through the interpreter. Other issues concerning the knowledge organization are the "granularity" of the knowledge stored in memory, and the number of different memories supported by the architecture and their relations. A cognitive robot needs both a *short-term memory* (for the storage of fast changing knowledge and short-lived goals and beliefs) and a *long-term memory* (for the storage of its generic concepts and competences that are relatively constant over time, although they can be updated via learning and instruction). In some cases the short-term memory of the architecture is treated as the active part of the long-term memory.

Knowledge employment   Knowledge can be employed in many ways ranging from low-level activities (e.g., recognition and decision making), to high-level activities (e.g., reflection and communication). A cognitive-robot architecture may support multi-step inference and problem-solving, using forward chaining, backward chaining, or their combination (known as means-ends analysis) [3]. Also, knowledge can be utilized using sequential or parallel processing suitably adopted for the cognitive architecture such that to cover both dynamically varying knowledge and parallel knowledge retrieval from long-term memory.

The capabilities of cognitive robot architectures include [1]:

Perception and situation assessment   A cognitive intelligent robot senses the world using several human-like sensing modalities (vision, hearing, tactile, etc.). Perception typically involves the integration of sensing signals from different sensors into a unique signal which is used for the assessment (or description) of the situation of the world in which the robot lives. Issues that must be faced in the perception process include the inaccuracy (uncertainty) of the sensors, which results in a partial picture of the robot's environment. This is more crucial if the environment itself is changing or involves uncertainties as it happens in all real-world (non-structured) environments. A cognitive intelligent robot must be able to interpret the situation of its broader environment (not only isolated objects or events) and, more importantly, to accurately predict future situations and events. This can be done by storing some mapping from a description of the current situation, and an action *onto* a description of the resulting situation. Learning can help very much in this prediction if supported by proper monitoring of the robot's and world's states.

Recognition and categorization   Recognition must not be restricted to static situations. For instance, a basketball player can recognize the execution of particular plays by the opposing team, which involves changing events. A soccer robot should have such a capability. Recognition contributes to object, situation, or event categorization by providing proper ways for representing patterns and situations in memory (via encoding general relations occurring across those situations). A complete cognitive architecture should include proper learning mechanisms to learn new patterns or categories from experience or instruction or to refine existing patterns as the case may be.

Reasoning and belief maintenance Reasoning is closely related to problem solving and planning. It draws logical conclusions from other assumptions or beliefs that the robotic agent already has. To be able to draw conclusions the robot architecture should be able to represent relationships among premises (e.g., via first-order logic, production rules, neural networks, Bayesian networks, etc.). Knowledge about reasoning may also be heuristic or approximate or fuzzy which can be processed by appropriate methods (heuristic reasoning, approximate reasoning, fuzzy reasoning). Deductive reasoning is the most widely used reasoning which allows one to combine specific or general premises to conclude logically new ones. Reasoning is used not only to derive new beliefs, but also to help the robot architecture in maintaining existing ones whenever required. For instance, because certain beliefs depend on others, the cognitive robot should trace the latter for determining whether it should continue to believe the former, abandon them, or generally change their confidence. Belief maintenance plays an important role in changing environments where the changes may be unexpected and non-predictable.

Decision making and selection among alternatives A cognitive robot should be able to make decisions and select among alternatives. Intelligent decisions are based on the recognition of an event, pattern, or situation. Most cognitive robot architectures possess both capabilities in a recognize-act cycle that exists in all cognitive processes. Decisions can be made within the architecture or, in case of more complex decisions, at a higher level enabled by the architecture. Of course, to be capable of decision making, a cognitive architecture must have some mechanism to represent internally or externally generated alternatives, and also a mechanism for selecting among these alternatives. This is typically done in two steps. The first step specifies if a selection is permissible (e.g., via pattern matching), and the second performs the actual selection among the alternatives. A cognitive robot should also be able to improve its decisions via learning or by revising the premises that led to a decision or changing the numerical functions employed at the conflict resolution step.

Planning and problem solving To achieve its goals in new situations, a cognitive intelligent robot should be able to generate plans and solve problems. Planning can only be performed if the cognitive robot possesses a model of the environment that can be used to predict the effects of its actions. This can be done by representing a plan as an ordered set of actions (at least partially), their predicted effects, and the way these effects make possible later actions. Plans that are valid for short time intervals can also be used profitably in fast changing environments or when the actions are abstract actions that can be expanded in several ways. A cognitive robot should also be capable of constructing plans from components available to memory that may refer to low-level sensor and motor actions or to more general plans including pre-stored sub-plans. If the planning refers to path planning, a cognitive mobile robot may create local path planning (in the nearby area of the robot's position) or global path planning in the entire space where the robot has to move. In practice, for successful motion both local and global path planning should be used. Problem solving is a process more general than planning since the robot needs to perform search through the application of actions or operators, instead of simply trying to develop a plan internally. Clearly, learning can contribute a lot to both planning and problem solving.

Communication and interaction   A cognitive robot should be able to acquire knowledge from another agent (human or robot) via proper com-munication and interaction. The transferred knowledge may contain any of the cognitive processes discussed so far (e.g., decisions, perceptions, actions, predictions, plans, inferences, and unexpected behaviors). The communication may be performed verbally or in written form using state-of-art conventions for semantics, syntax, etc. Of particular popularity are conversational dialogues in natural language which requires the robot to be able to understand natural language combined with the ability of turn taking in cooperation with the other agent.

Execution and action capability   A cognitive robot should be able to represent, store and execute motor actions in the environment. These actions and skills may be encoded via primitive or component actions possibly including multi-step skills or procedures (encoded in the form of plans retrieved from the memory). Both open-loop and closed-loop strategies for execution based on the feedback information provided by the sensors of the robot should be embedded. Again, learning about motion skills and execution policies can be used in several forms as explained in previous capabilities (learning through observation, learning through imitation, learning through instruction, etc.), and discussed below.

Learning, memorizing and reflection   Learning is a basic component of all architectural cognition capabilities discussed so far. Typically, learning provides generalizations, beyond specific beliefs and events. This generaliza-tion can be performed at storage time and introduce the generalized knowledge into the memory, but can also occur at retrieval time via analogical or case-based reasoning. In many cognitive architectures, learning is performed in an automated manner which cannot be inspected or consciously controlled. However, even in this automated learning case learning can be made in a more deliberative way using meta-reasoning. Learning data may be acquired from many sources such as observation of another agent, practice of known skills or the robot's own problem-solving capacity. A prerequisite of learning is the capability to remember (memorize), i.e., to encode and store the results of cognitive processes and retrieve or access them at a later time. Reflection is the capability to access traces of cognitive activity such as recent mental structures (still available) or older structures that may be retrieved from the episodic memory. Examples of reflective activity include the explanation or justification of the robot's plans, decisions, inferences or actions through the steps followed to achieve them. Another way to reflection is the meta-reasoning about other cognitive processes that can be used by the architecture as explanation with emphasis on their generation. Reflective processes are themselves subject to reflection in case they have access to their own cognitive traces.

9.3 Taxonomies of Cognitive Intelligent Robot Architectures

Two popular complementary taxonomies of cognitive/intelligent architectures are the following [4–6]:

Taxonomy 1

- Hierarchical architectures
- Behavior-based architectures
- Hybrid architectures

Taxonomy 2

- Cognivist/symbolic architectures
- Emergent architectures
- Hybrid cognivist-emergent architectures

A discussion of these taxonomies follows.

9.3.1 Taxonomy 1: Hierarchical, Behavior-based, and Hybrid Architectures

Hierarchical architectures   These architectures follow the top-down style which highlights the superiority of high-level control, and restricts low-level horizontal (flat) communications. A representative hierarchical cognitive, control architecture involves three levels each one containing one or more layers, similarly to human control mode of interaction between director (supervisor) with his/her subordinate employees. These levels are as shown in Figure 9.1 [7, 8]:

- Organization level
- Coordination level
- Execution level

This architecture is also known as *supervisory intelligent control architecture* and has been thoroughly studied by George Saridis [9]. Using this control architecture he formulated the so-called *"principle of increasing precision with decreasing intelligence"* (**IPDI**) which is a manifestation of the human organizational pyramid. Specifically, as we proceed from the upper (organization) level to the lowest (execution) level the precision of control is increased but the degree of intelligence/cognition ability is decreased. The organization level is designed to organize a sequence of high-level skills/actions, and the execution (actuation) level performs the proper control functions for achieving the desired goal(s). The coordination level serves as an interface between the organization and execution levels. The analytic theory developed by Saridis assigns analytical models to the various levels in the control hierarchy, and has been applied to particular paradigms of cognitive and autonomous robotic systems [9]. Clearly, hierarchical architectures have their modules and components disposed in a certain *a priori* structure that involves several levels and layers.

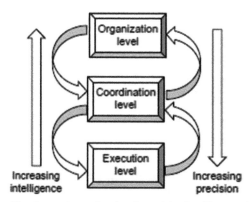

Figure 9.1   The three primary levels of cognitive/intelligent robot control.

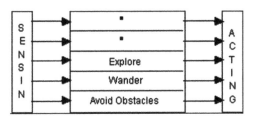

Figure 9.2   The subsumption architectural model.

Behavior-based architectures   These architectures are based on the concept of *"agent"* and can be implemented using knowledge-based (symbolic) systems, neural (non-symbolic) systems, and fuzzy or neurofuzzy structures [4, 10, 11]. In a behavior-based architecture each module or level directly generates some part of the cognitive robot's behavior, and incorporates its own perception, modeling and planning requirements (e.g., one module may know how to move avoiding obstacles and stairs, and another how to build a map of the environment, and so on). The two earliest generic behavior-based architectures are the *subsumption* architecture developed by Brooks [12, 13] and the *motor-schema* architecture developed by Arkin [4, 14, 15]. The subsumption architecture follows the decomposition of the behavior paradigm (Figure 9.2), and was firstly used in the autonomous mobile robot Shakey.

The motor schemas architecture was more strongly motivated by biological sciences and uses the Kantian theory of schemas which represents a means by which understanding is able to categorize sensory perception in the process of realizing knowledge of experience. A schema is defined in several ways, for example: (i) as a pattern of action or a pattern for action, (ii) a perceptual entity corresponding to a mental entity, or (iii) a functional unit that receives special information, predicts a possible perceptual content, and matches itself to the perceived information.

A convenient working definition is the following [11]:

"A schema is the fundamental entity of behavior from which complex actions can be constructed, and which consists of the knowledge how to act or perceive, as well as the computational process by which it is enacted".

Motor schema analysis and design of behavior-based systems possesses the following capabilities:

- It can explain motor behavior in terms of the concurrent control of several different activities.
- It can store both how to react and how to realize this reaction.
- It can be used as a distributed model of computation.

- It provides a language for connecting action and perception.
- It provides a learning approach via schema elicitation and schema tuning.
- It can explain the intelligence functions of robotic systems.

Motor schema behaviors are relatively large grain abstractions, which can be used in a wide class of cases. Typically, these behaviors have internal parameters which offer extra flexibility in their use. Associated with each *motor schema* there is an embedded *perceptual schema* which gives the world specific for that particular behavior and is capable of providing suitable stimuli.

Hybrid hierarchical and behavior-based architectures   Hierarchical architectures are of the deliberative type where the policy of access to the environment is carried out by the organizer (or planner) which is actually a specialized arbiter. In reactive behavior-based control system a permanent interaction with the environment is maintained and all of its modules are connected to the sensors and/or actuators of concern.

Three ways in which *planning (deliberative)* and *reactive* behavior can be merged are [4]:

- Hierarchical integration of planning and reaction (Figure 9.3a)
- Planning to guide reaction, i.e., permitting planning to select and set parameters for the reactive control (Figure 9.3b)
- Coupled planning-reacting, where these two concurrent activities, each guides the other (Figure 9.3c).

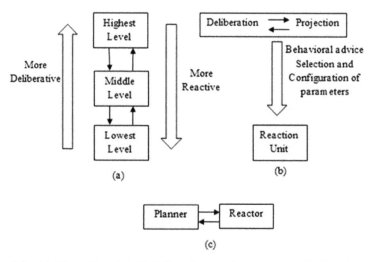

Figure 9.3   (a) Hierarchical hybrid deliberative-reactive structure, (b) Planning to guide reaction scheme, (c) Coupled planning and reacting scheme.

One of the first robotic control schemes that were designed using the hybrid deliberative (hierarchical) and reactive (schema-based) architecture is the *Autonomous Robot Architecture* (AuRA) [4]. AuRA incorporated a traditional planner that could reason over a modular and flexible behavior-based control system.

9.3.2 Taxonomy 2: Symbolic, Emergent, and Hybrid Architectures

This architectural taxonomy was examined to some extend in Chapter 4 (Section 4.6). Here, we will discuss this taxonomy on the basis of two primary cognitive capabilities, namely *memory* and *learning* [6].

Memory has been considered in the literature from several points of view. Several types of memory can serve for the storage of knowledge about the world and oneself, and for the storage of the current episode of activity, whereas learning is the principal process that shapes this knowledge. Learning and memory together constitute the fundamental aspects of cognition on which higher-order processes and cognition/intelligence capabilities, including planning, deliberative reasoning, and self-control, are built. The memory organization depends on the adopted knowledge representation formalisms. This taxonomy (cognivist or symbolic, emergent, and hybrid architecture) is pictorially depicted in Figure 9.4 where the principal types of memory and learning are included [6].

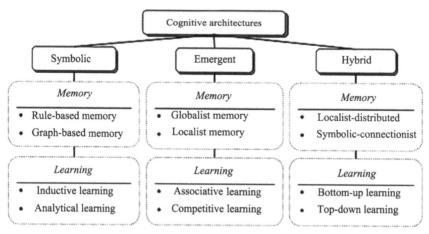

Figure 9.4 Pictorial representation of the cognitive architectures taxonomy.

Source: ResearchGate DuchTaxonomy-CogArch.jpg.

Symbolic architectures use high-level symbols for knowledge representation and processing, whereas emergent architectures employ activation signals which pass through neurons, and a bottom-up process based on the emergent features of self-organization and association. Hybrid architectures employ various combinations of cognivist and enactive characteristics. In most cognivist architectures a centralized control style is employed over the information flow from sensory inputs via memory to activators' outputs. Rule-based memory of perception-action cycles, include human-like skills of logical reasoning. Graph-based representations are employed for encoding various graph structures (semantic nets, conceptual graphs, etc.) with *nodes* standing for symbolic entities and their attributes, and *edges* standing for their relationships. As discussed in many occasions of the book, learning can have several styles (analytical learning, inductive learning, explanation-based learning, analogical learning, neural network learning, reinforcement learning, imitation learning, learning through observation, etc.).

9.4 The Meta-Cognition Concept in Robotic Systems Architecture

The *meta-cognition* concept was inspired from the *metaphysics* (or *ontology*) philosophical concept according to which an attempt is made to list, in order of priority, the various kinds of entities and processes that make up the universe (physical objects, properties, attributes, events, bodies, etc.). A design option in hierarchical architectures is to incorporate a sort of *"alarm agent"* which is always operating and monitors internal or external processes and is able to detect quickly situations that need emergency control actions possibly changing the behavior of many other agents (modules) of the system. Since it is possible for an alarm mechanism to produce false alarm signals, a *meta-alarm* control function is needed to determine when alarm signals are allowed to interrupt ongoing processes (typically using varying attention filter thresholds).

Architectures with *meta-cognition*, or, more generally, *meta-management* functions can support various forms of mental states [16, 17]. In humans, these meta-management level functions develop over many years, leading to high-level control mechanisms that defined the *"personality"* of human beings (called *personae*). A hierarchical cognitive robotic architecture that uses reflective meta-management processes for the higher-level control at the deliberative layer, the reactive layer, the memory subsystem, and the alarm mechanism is the *H-Cog Aff* architecture [17–19]. It is remarked that a dedicated meta-level module may be capable of improving specific modules independently without any undesired interaction. It can also find out when no existing module is able to sufficiently perform a certain task, and generate a new module that can perform the task under proper applicability conditions. Meta-level decisions may themselves present complex problems, and their switching operations may be monitored and corrected by higher-level controllers. In general, such a meta-management control architecture may contain many meta-control layers.

This meta-control philosophy is pictorially illustrated in Figure 9.5 [16].

The environment (ground level) may involve complex physical and social environments that can be perceived by sensors and regulated by proper effectors. The sensor/effector-environment coupling is continuous (i.e., it does not alternate between being *ON* or *OFF*), the sensor signals are analyzed by central processors which also perform the action selection and decide what to do. This means that instead of the conventional action-perception cycle the architecture needed must involve concurrent processes of several forms that can interact with one another. These enduring (concurrent) processes take place at different levels of granularity, on different time scales (changing continuously or discretely), and perform different functions. Clearly, this structure has implications for the knowledge representation, reasoning, computation and conflict resolution, and needs very high computational power. Summarizing, instead of designing a control scheme that can deal with all possible situations, we use a new mechanism (meta-level control mechanism) that monitors the system performance, and modifies the behavior *on-line* by changing parameters or altering the module concerned such that to modify future behaviors (via self-correction, learning, etc.).

Meta-management actually occurs in consciousness (including self-awareness, and external "self-consciousness", i.e., things that other agents might know about the individual agent). Meta-management can be used for the following architectural aspects [16]:

Meta-semantic competence This is needed when a system uses information about information or information of entities that acquire or about the representation of things that it represents.

Meta-theory development This refers to the meta-management capability to discover the need to modify or extend existing theories about how events occur and are used in the environment. Such a modification is required whenever some things perceived in the environment are in conflict with predictions that are based on current theories, and can be done either by *abduction* (i.e., by deriving a new theory using available concepts) or by using new concepts that refer to hypothesized (non-observed) properties that explain behavior.

Uncertainty meta-management   Humans have the ability to predict changing epistemic affordances using probability distributions or fuzzy

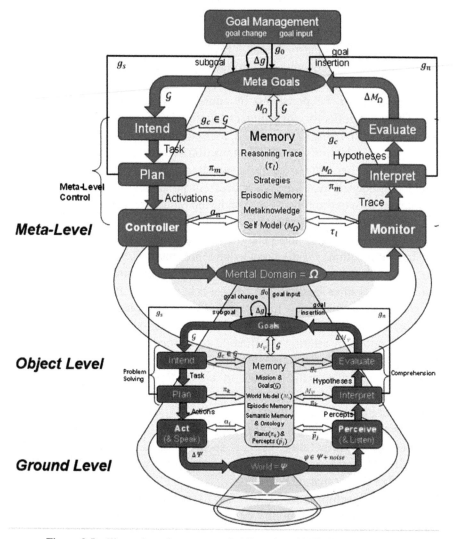

Figure 9.5   Illustration of meta-control philosophy of MIDCA architecture.

Source: www.wright.edu/michael.cox/

membership functions that express the involved uncertainty. The need to use probabilistic or fuzzy reasoning can be avoided by using meta-knowledge that there are *regions of certainty* [**ROC**] (definitely **YES** and definitely **NO**) with intermediate *regions of uncertainty* or fuzziness [**ROU**]. Using epistemic affordances an agent (human, robot) can move out of the ROU into ROC, e.g., by changing the direction of gaze, changing view point, altering the direction of motion, or rotating an object, etc.

Meta-cognitive mathematics   In mathematics there are *truths (invariant features)* about topology and geometry. Meta-cognitive reflection on invariant mathematical features of what is perceived is based on Immanuel Kant's philosophical argument that mathematical knowledge is *synthetic* and *non-empirical*. As argued by Sloman [20, Ch. 8] robots can learn mathematics as a consequence of reflecting on things they learn about actions and processes in the environment. Many of these discoveries refer primarily to features of static geometric structures (e.g., the angles of a triangle sum up to 180°) or to invariant patterns in sets of sentences that lead to logical discoveries (Aristotelian, Boolean, etc.). Sloman [16, 20] argues that logical knowledge is the outcome of using meta-cognition mechanisms involved in discovering new mathematical knowledge.

9.5 Some Representative Cognitive-Intelligent Robot Control Architectures

Here, a few representative cognitive/intelligent control architectures will be outlined. Other architectures are described in survey/state-of-art papers (e.g., [5, 6, 21, 22].

9.5.1 Hierarchical and Behavior-based Architectures

A social robot hierarchical cognitive architecture   This architecture is one of the earliest hierarchical (layered) architectures developed at MIT by Rodney Brooks and his associates for the humanoid socialized robots **Gog** and **Kismet** [12, 23, 24]. In this architecture *perception* is distinguished in *high-level* perception (gesture recognition, attention state recognition, recognition of the self and the other, face and eye detection, speech recognition) and *low-level* perception (auditory feature extraction, tactile/and kinesthetic sensing, motion detection, obstacle detection, etc.). Gesture recognition is used for performing shared attention functions, e.g., learning from an instructor by attending the same objects that he/she looks, etc. Socialized robots need also a *motivation* *system* which enables the robot to acquire social inputs and understand human cues and emotions, and handle its environment. The motivation system is complemented by an *attention module* which enables habituation, and integration of low-level motivation effects. All these capabilities provide proper signals to the motor system through proper hardware and software interfaces. Figure 9.6a shows a possible way of interlinking the subsystems of the cognitive robot in a 3-level architecture, and Figure 9.6b gives another architecture that involves five parts.

In Figure 9.6a the low-level perception signals are sent to the attention system, which selects the relevant ones at that time, and directs the robot's attention and gaze towards them. The motivation system communicates

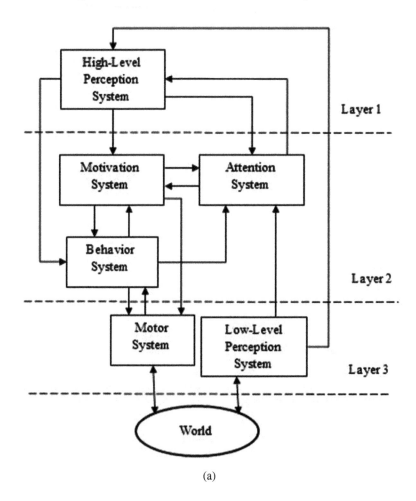

(a)

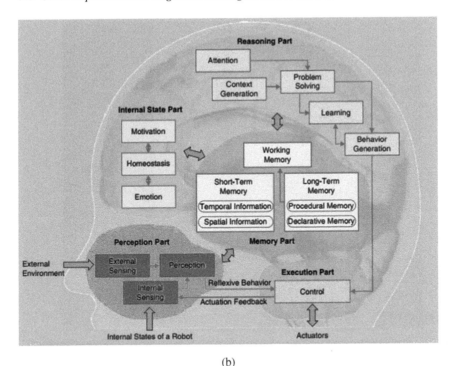

(b)

Figure 9.6 (a) A 3-layered architecture for interlinking the six systems of a cognitive humanoid socialized robot. (b) An alternative behavior-based humanoid cognitive control architecture composed of five self-explained parts (perception part, memory part, internal state part, reasoning part, and execution part).

Source: http://rit.kaist.ac.kr/home/Default/Research3?action=AttachFile&do=get&target=ioa.png

bilaterally with the attention system, and involves the robot's basic "emotions" and "drives". The motor system receives commands from the motivation and behavior systems in order to perform the blending and sequencing of the elementary actions from the corresponding socialized motor/actuation units.

The LAAS hierarchical behavior-based architecture This architecture involves two levels (decisional level, functional level), as shown in Figure 9.7 [25, 26]. The highest decisional level involves the mission planner, and the intermediate layer involves the task refinement (task supervision).

The *functional modules* involve build-in modules for modeling, monitoring, reflex actions, servo control, image processing, etc., which communicate without any fixed hierarchy. The *Executive system* controls and coordinates

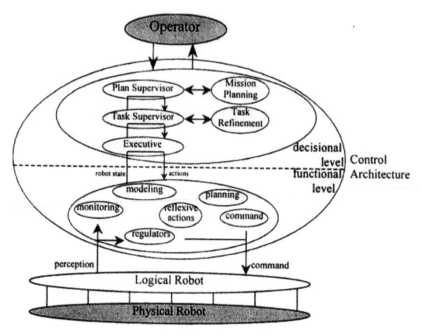

Figure 9.7   The LAAS hierarchical architecture.

dynamically the execution of the functions, distributed in the modules, according to the task requirements specified by the *Task decision system*. The highest decision level including the capacities for generating and organizing the task sequence plan and supervising its execution, and at the same time being capable to react to events from the other (lower) levels. It may involve several layers which integrate deliberation and reaction, according to the application at hand. Actually, the LAAS architecture is an example of *hybrid deliberative and reactive* behavioral architecture. The hierarchy of levels implements a vertical decomposition, while under certain conditions, the modules can operate in parallel which typifies an horizontal (lateral) decomposition of reactive subsystems. General criteria satisfied by the LAAS architecture are: reactivity, flexibility, re-configurability, autonomy, and programmability. A special temporal planner (**IxTeT**) is used to reason about the symbolic or numerical relations, which produces the plan in the form of a partially ordered set of tasks in compatibility to the time constraints such that minimal and maximal bounds on the durations of the execution are defined and used for synchronization with external events.

9.5.2 Cognivist Emergent and Hybrid Architectures

The SOAR cognivist architecture This is a classical example of symbolic (rule-based) cognitive architecture designed to model general intelligence [27]. The name **SOAR** is the acronym of **S**tate, **O**perator **A**nd **R**esult. SOAR stores the knowledge in the form of production rules, arranged in terms of operators that act in the problem space which is the set of states that represent the task at hand. Its primary learning mechanism is a type of **E**xplanation **B**ased **L**earning (**EBL**), called *chunking* [28], formulating rules and macro-operations from problem solving traces. Newer extensions of SOAR architecture include reinforcement learning to adjust the priority parameters of operators, semantic learning and episodic learning to retain history of system evolution [29, 30] Figure 9.8 shows a non-detailed diagram of the SOAR architecture which involves the chunking mechanism, the

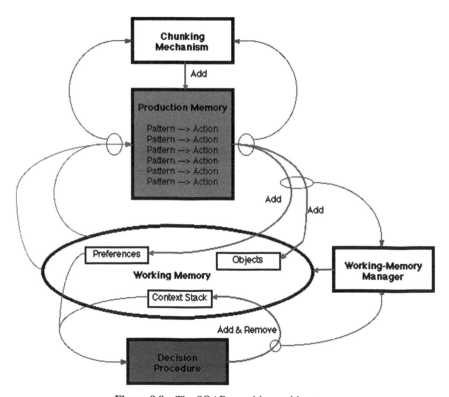

Figure 9.8   The SOAR cognitive architecture.

Source: http://www.it.bton.ac.uk/staff/rng/teaching/notes/Soar1.gif

production memory, the working memory (attributes, values, preferences, objects), the working memory manager, and the decision procedure blocks. The knowledge representation is fine-grained and the sub-goal is universal. SOAR architecture has exhibited a variety of high-level mental functions in real-time distributed environments for solving complex problems, understanding natural language and making complex plans with its ability to process large and complex rule sets.

The ICARUS cognivist architecture　This is an integrated cognivist architecture for physical agents (robots, etc.) with knowledge specified in terms of reactive skills [31]. The cognition is grounded in perception-action style, and the concepts and skills are distinct cognitive structures that are acquired cumulatively. Long-term and short-term structures have a strong correspondence, and the symbolic cognitive structures are modulated using numerical functions. The long-term memory is organized in hierarchical form with the conceptual memory directing bottom-up percept-driven reasoning, means-ends analysis, and skill memory controlling in a top-down way. The acquisition of knowledge in ICARUS is performed via hierarchical, incremental reinforcement learning propagating reward values backwards over time.

A non-detailed diagram of the ICARUS architecture is shown in Figure 9.9.

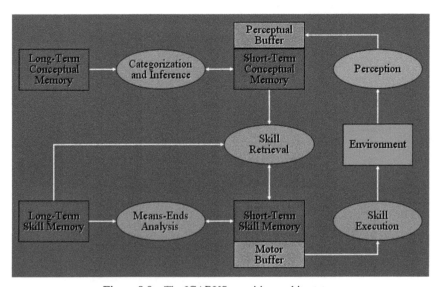

Figure 9.9　The ICARUS cognitive architecture.

Source: http://www.isle.org/˜langley/talks/icarus.6.04.ppt

When ICARUS can find a path that terminates in a primitive skill with executable actions, it applies these actions to affect the environment. This leads to new percepts, changes in beliefs and reactive execution of additional skill paths to achieve the agent's goals. But if ICARUS cannot find any applicable path through the skill hierarchy which corresponds to a top-level goal, it proceeds to problem solving using a kind of means-ends reasoning [32].

The Global workspace emergent architecture In this architecture, anticipation and planning are achieved through internal simulation. The action selection (internal and external) is mediated by affect. The global workspace model is employed where parallelism is a basic feature of the architecture (i.e., not an implementation issue). The knowledge representation is analogical which is characterized by a small semantic gap and more convenient grounding. A top-level sketch of the *GW architecture* is given in Figure 9.10 [33, 34].

In Figure 9.10 there are two interacting sensorimotor control loops. The first-order (external) loop is purely reactive, and specifies an immediate response to the current situation without the participation of cognition. However, these unmediated motor responses are subject to a veto by the *"basal ganglia* (**BG**) *analog"*, which also carries out salience-based action selection. The higher-order loop which modifies the behavior of the first-order (inner) loop is closed through **BG** and the *amygdala analog* (**Am**).

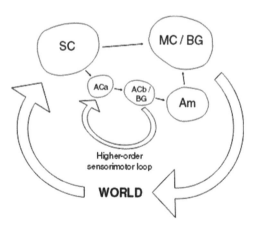

Figure 9.10 The top-level schematic of the global workspace architecture (**SC** = Sensory Cortex, **MC** = Motor Cortex, **AC** = Association Cortex, **BG** = Basal Ganglia/action selection, **Am** = Amygdala/affect). Courtesy of Murray Shanaham [33, 34].

Source: http://www.doc.ic.ac.uk/~mpsha/ShanahanConCog06.pdf

This is performed by adjusting the salience of presently executable actions. In some situations, this adjustment has no effect, and in other cases it will result in a new action so becoming the most salient. The global workspace architecture is based on Baars, Edelman and Gally *global workspace theory* (**GWT**) which "hypothesizes that a number of brain components constitute and integrative workspace that serves to reconcile the narrow momentary capacity of conscious contents with a widespread recruitment of unconscious brain functions, including long-term memory" [35]. Furthermore, Baars argues that the architecture of the brain can be best understood as a *working theater*, which is pictorially depicted in Figure 9.11 [36].

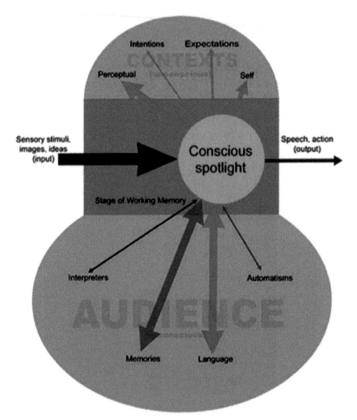

Figure 9.11   The working theater representation of the brain. Courtesy of Bernard Baars [35, 36].

Source: http://bernardbaars.pbworks.com/f/1205970309/carpenter-gwt-theater.jpg

The narrow limits of consciousness capacity include:

- The working memory.
- The selectivity (i.e., consciousness is limited to only one dense stream of input).

Unconsciousness includes a vast collection of unconscious brain processes, namely:

- The mental lexicon, the meaning, and the grammar.
- Autobiographical memory.

Figure. 9.12 illustrates the broadcasting of consciousness model as suggested by GWT (i.e., consciousness is the result of highly coordinated, widespread activity of the brain).

- *Conscious senses* occur when signals are broadcast to a wide network of neurons that covers much of the cortex.
- *Unconscious senses* occur when signals stay localized.

The LIDA hybrid cognivist/emergent architecture   The name of this architecture is the acronym of "**L**earning **I**ntelligent **D**istribution **A**gent". It combines symbolic and connectionist knowledge representation and processing. It is partly composed of entities at a relatively high level of abstraction, such as behaviors, message-type nodes, emotions, etc., and partly of low-level *codelets* (small pieces of code). LIDA is both a computational architecture and a working model of the brain [37, 38]. All symbols of LIDA are grounded in the physical world, in the sense of Brooks and Stein [39]. LIDA can perceive both endogenously and exogenously. Its perceptual knowledge-base, called *perceptual associative memory,* takes the form of a semantic net, with activation, called *"slipnet"*. Nodes of the slipnet represent the agent's perceptual symbols

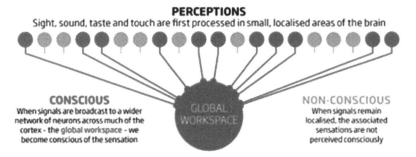

Figure 9.12   The consciousness broadcasting model. Courtesy of Bernard Baars [35, 36].

Source: http://universe-review.ca/I10-80-globalws.jpg

referring to individuals, categories, and possibly to higher-level concepts and ideas. LIDA has separate modules for perception, working memory, semantic memory, episodic memory, emotions, action selection, expectation, learning procedural tasks from experience, deliberation, negotiation, problem solving, constraint satisfaction, meta cognition, and conscious-like behavior.

The LIDA architecture is primarily based on the Global Workspace theory discussed above, and embeds a number of psychological and neuro-psychological theories. Consciousness is modeled by the multitude of codelets, some of which activate nodes in the global workspace. Codelets scan the inputs on-line to see if there is anything new.

To achieve its goals, LIDA employs several subunits namely:

- Coalition manager
- Spotlight controller
- Broadcast manager
- A number of codelets

The perception is performed by the codelets that act on the inputs. The coalition manager regulates the coalitions, i.e., the attention codelet with its associated codelets. The strength of the coalition indicates how well the codelets match the input to the conditions they are to identify. The spotlight controller determines which of the coalitions are to be elevated to the global workspace. Typically, the spotlight controller selects the coalition with the highest strength. The coalition elevated to the global workspace is broadcast to the entire system by the broadcast manager. The action selection is performed by behavior codelets where each behavior is of the form of an IF–THEN production rule that has premises and associated actions (conclusions) which take into account the strength of the premises.

The above operations are performed in a cognitive cycle (called the *LIDA cognitive cycle*) which involves nine steps grouped in three phases as follows (Figure 9.13).

The understanding phase (steps 1, 2 and 3): Here the incoming stimuli are scanned by independent parallel codelets and activate low-level feature detectors in *Sensory Memory*. The corresponding output engages the *perceptual associative memory* where more abstract entities are stored, such as objects, events, actions, etc. The resulting percept enters the workspace where it cues both *episodic (transient) memory* and *declarative memory* producing local associations. These associations are combined with percept to produce a current situational model which represents what is happening right know.

The attention/consciousness phase (steps 4 and 5): This phase forms coalitions of the most salient parts of the current situational model which

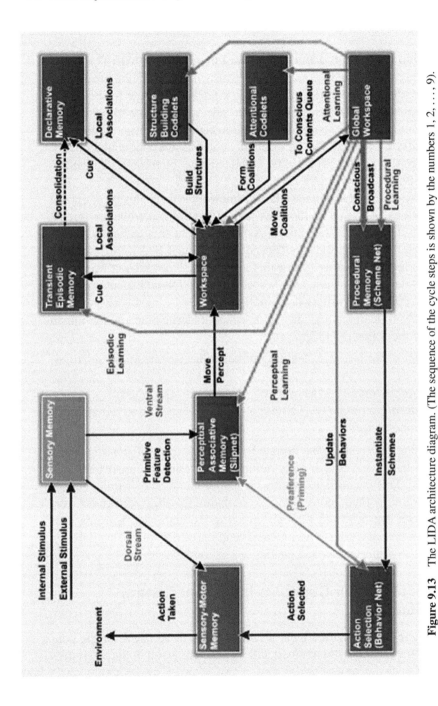

Figure 9.13 The LIDA architecture diagram. (The sequence of the cycle steps is shown by the numbers 1, 2, . . . , 9).

Source: http://www.garfixia.nl/l/library/download/urn:uuid:d084cf40-8eb4-4f5e-bad3-8252b542c99f/ IDA.jpg?height=462&width=756&ext=.jpg (Also: LIDA Tutorial.pdf)

is then compete for attention (i.e., for a position in the current conscious contents).

The action selection and learning phase (steps 6 through 9): This phase is initiated by the alternative actions that are instantiated from the *procedural memory* and are sent to the *action selection model* to compete for being the behavior chosen by this cognitive cycle. The selected behavior stimulates the *sensor-motor memory* to generate an execution algorithm. This completes the LIDA cognitive cycle. LIDA's *"life"* actually consists of a continual sequence of these cognitive cycles. The LIDA architecture is pictorially depicted in Figure 9.13 [40].

One of the main features of the LIDA framework is the inclusion of default implementations of its major modules and several abstract classes for the generic parts of the architecture on LIDA of other agent architectures [56, 57]. The original IDA computational model and its extension LIDA were developed at Stan Franklins "Conscious" Software Group at the University of Memphis. IDA was not designed to learn, and its software code (about a quarter million lines of Java code) was entirely hand crafted.

The modes of learning added to IDA, which underlie much of the human learning, are the following [37]:

- *Perceptual learning* (i.e., learning of new objects, categories, relations, etc.).
- *Episodic learning* (i.e., learning of what, where, and when).
- *Procedural learning* (i.e., learning new actions and action sequences).
- *Attentional learning*.

These learning procedures are initiated by the contents of consciousness during the LIDA cycle, and is fully automatic (i.e., unconscious) from that point onward. This means that actually conscious attention is learning. Thus, volitional learning (and more complex problem solving) performed over multiple cycles are achieved through directing the automatic learning processes. LIDA possesses the consciousness features of difference, availability, phenomenal/structural unity, and qualia.

9.6 A List of Further Cognitive-Intelligent Control Architectures

Here, ten more cognitive/intelligent architectures will be listed for the convenience of the reader, without any details. These are given in the references cited.

Architecture 1: TCA architecture This architecture (Task Control Architecture) provides framework for controlling distributed robot systems, with the aid of mechanisms that support distributed communications, task decomposition, resource management, execution monitoring, and error recovery. TCA involves a number of task-specific modules and a general purpose reusable central control module [41].

Architecture 2: SSS architecture This architecture involves three layers, namely: a servo-layer (the bottom layer), a "subsumption layer (middle layer), and a symbolic layer (top layer) such that the advantages of each technique can be fully exploited. This is achieved via a synergy interface between the subsystems involving situation recognizers that bridge the gap between the servo and subsumption layers, and event detectors that link the subsumption and symbolic layer [42].

Architecture 3: SRA architecture This social robot architecture provides a framework for explicit human-robot social interaction. SRA adopts the stance of an ego-centric robot control strategy, developed to socially extend to the agents the robot engages with, and is intended to be a more generic control structure for robots that can exhibit social behavior [43].

Architecture 4: EPIC-Executive Process Interactive Control This is a cognitive architecture encompassing many aspects of human behavior (perceptual, cognitive, motor actions), using serial interconnected processors working in parallel. It can perform human computer interaction for practical control purposes. The system control is implemented via production rules for the cognitive processor, and a set of perceptual (visual, auditory, tactile) and motor processors operating on symbolically coded features [44]. EPIC, combined with SOAR, has been used for simulating traffic control [45].

Architecture 5: Human – Aware Cognitive Architecture This architecture integrates concurrent processes and interacting subsystems that behave according to different temporal properties, and implements the cognitive capabilities of interpretation, decision making, learning and communication. It allows the robot to execute its tasks and also to produce behaviors that support its commitment *vis-à-vis* its human partner, and to interpret human behaviors and intentions [46].

Architecture 6: CogAff architecture This architecture merges cognitive and affective components. It has a hierarchical structure with information flowing through perceptual, central and motor levels which have several layers performing tasks of varying abstractness, using several mechanisms and representations. Principal modules of the architecture are the module of alarms and the meta-management module that performs the functions discussed in

Section 9.4. An enhanced version of this architecture, the **H-CogAff** *(Human-like CogAff)* involves additional components and features that perform motive processing and "what if" reasoning in the deliberative layer, below the meta-management layer. CogAff is also referred to as a schema rather than an architecture, because it is not necessary every component specified in it to be included in every architecture to which is relevant [19, 47].

Architecture 7: Agent-based cognitive robot architecture   In this architecture the cognitive layer can be programmed by means of the agent programming language **GOAL**. This language is suitable for creating cognitive robotic agents including symbolic knowledge representation, deliberation via modular high-level action selection, and supports multiple declarative goals. It supports for a clear separation of symbolic and sub-symbolic reasoning and processing [48].

Architecture 8: DUAL architecture   This is a hybrid multi-agent general-purpose architecture that supports dynamic emergent computation. This is done with a unified description of mental representation, memory structures, and processing mechanisms performed by small interacting micro-agents. Due to that a central controller is lacking the system is continuously changing following the changes of the environment. The agents of the architecture interact to form larger complexes, coalitions, and schemas. The symbolic representation of facts is performed by microframes, with activation levels of these facts working in a particular context of network interconnections [49].

Architecture 9: NARS: Non-Axiomatic Reasoning System   This is a reasoning system based on a knowledge representation language which involves an experience-grounded semantics of the language, a set of inference rules, a memory structure, and a control mechanism that carries out various high-level cognitive tasks as different aspects of the same underlying processes. The non-axiomatic logic is convenient for adaptation with insufficient knowledge and resources that operate on patterns having "truth-values" evaluated according to the systems "experience" with the use of these patterns [50].

Architecture 10: PRODIGY architecture   This architecture works with two types of knowledge, namely *domain rules* (that encode the conditions under which actions have certain effects specified by the addition or deletion of first-order expressions), and *control rules* (that determine the conditions under which the selection, rejection or preference of a given operator should be made). PRODIGY performs search within a problem space to achieve the desired goal(s), and uses the control rules to select an operator, binding set, state, or goal. On the basis of its problem-solving experience, PRODIGY constructs EBL rules, and has a mechanism which analyzes problem-solving traces providing the reasons of success or failure. Also, it includes separate modules to control search by analogy with existing solutions, observed current solutions, or experimentation [51, 52] (see also PRODIGY 4.0: The Manual and Tutorial).

9.7 Evaluation and Validation of Cognitive-Intelligent Architectures

As it is clear from the exposition of the architectural examples offered in Sections 9.5 and 9.6, although the architectures are designed and implemented via essentially the same information processing tools, they are based on different conceptual models or sub-models of cognition and intelligence, and exhibit several differences in their capacities, features and effectiveness. Therefore, each one of them has to be evaluated and validated (and possibly compared with the other architectures developed for the same application repertory).

This evaluation can be performed by employing several distinct criteria, including criteria referring to their ability to explain psychological and mental phenomena. A list of criteria synthesized from [1, 21, 22, 53, 54] is the following:

- Generality and portability.
- Extensibility and scalability.
- Autonomy and adaptability.
- Reusability and repeatability.
- Robustness and consistent behavior.
- Reactivity and safety.
- Optimality and improvability.
- Documentation.

Of course it should be noted that no single architecture satisfies the entire set of criteria but only various subsets of this list. A brief discussion of them is as follows.

Generality and portability   Generality is required because cognitive architectures aim to support general (generic) intelligence. Generality can be evaluated and validated by applying the architecture to a diversity of tasks, environments, and situations. Portability is the capability of an architecture including software and hardware (sensors, actuators, processors, etc.) to be ported easily in other software/hardware frameworks.

Extensibility and scalability   Extensibility refers to the capability to add new modules to the system, and to the ability to be used for performing new tasks in response to goals or other external commands from other agents. Scalability refers to the ability of an architecture to deal with situations of different difficulty and complexity, and can be validated on a *live-when needed* basis.

Autonomy and adaptability   Autonomy and adaptability refer to the capability of an architecture to carry out its actions and apply its skills (refining or modifying its own behavior) according to its own goals and to the execution context as perceived by its sensors. Autonomy of an architecture can be tested by presenting agents with high-level tasks which for their success need autonomous decision making and knowledge acquisition about the application. Adaptability can be evaluated by checking the capability of an architecture to adapt to new situations and change its performance (rules, actions, skills, etc.) such that to face the internal or external (structural, sensing, or command) variations.

Reusability and repeatability   These criteria refer mainly to the software and computational facilities of the architecture. Reusing software components, structure, and framework and software patterns (conceptual, analysis, or design patterns) can speed-up the architecture's software implementation. Software repeatability means that running the same program on the same input gives the same result.

Robustness and consistent behavior   A specific function has well defined operating conditions. Robustness is the capability of the architecture to face situations out of the scope of these conditions. Typically this is done by exploiting the redundancy of the architecture functions. For consistent behavior the actions and reactions of the robot must be guided by the objectives of its tasks.

Reactivity and safety   An architecture has to take into account events and situations with time limits compatible with the correct and efficient achievement of the robot's goals, including its safety which is a permanent goal. Reactivity can be measured in terms of the speed with which the architecture (robot) responds to unpredictable events or situations, or by using the probability of responding on a given 'percept-act cycle'.

Optimality and improvability   Optimality is defined in terms of some performance index (typically numeric) which takes values over a space of behaviors. The optimal behavior corresponds to the best (maximum or minimum) value of this performance index. The degree of optimality can be measured by determining the fraction of times the architecture's performance is optimal across several decision cycles or by finding the average ratio of actual-to-optimal value achieved over time. The improvability of an architecture is its capability to improve over time. It can be measured using the robot ability to perform tasks that it could not achieve before the addition of more knowledge.

Documentation   When a cognitive robot architecture is to be used nonlocally (by other people than its developers) it should be accompanied with proper and rigorous documentation. This documentation includes:

- The philosophy of the architecture
- An accurate guide for the programmer (and user)
- A reference manual
- Code documentation
- Proper conceptual and methodological tutorials for the convenience of the user (printed and/or web-based)

Validation of an architecture involves *testing* and *formal verification* [54]. Testing involves both unit/module/component testing and overall system testing, where the former testing should be made before the entire system testing. Clearly, the response of a unit/module/component is typically affected by the response of other units or components. Therefore, testers usually need to remove components and replace them with others functionally equivalent but simpler, in order to see how the component/unit under test behaves in that context. Testing is a tedious process, but it can be accommodated in the architecture by using anonymous publish-subscribe messaging [55]. For the validation, in many cases the robot is replaced by simulation. However, in these cases the speed difference between the real and simulated robot has to be taken into account so as to assure that the relative timing of events is the same in both of them.

A comparative review of 20 cognitive architectures was provided in [21, 22], based on the following seven criteria:

- Embodiment
- Perception
- Action
- Anticipation
- Adaptation
- Motivation
- Autonomy

These architectures were grouped in the three categories: *cognivist, emergent,* and *hybrid.* The evaluation/comparison results are summarized in Table 9.1.

Table 9.1 Comparison of cognivist, emergent, and hybrid cognitive architectures (Adapted from [21, 22])

Cognitive Architecture	Embodiment	Perception	Action	Anticipation	Adaptation	Motivation	Autonomy
Cognivist							
SOAR	N	N	N	+	+	N	N
EPIC	N	+	+	+	N	N	N
ACT-R	N	+	+	+	+	N	N
ICARUS	N	+	Y	+	Y	N	N
ADAPT	Y	Y	Y	+	+	N	N
GLAIR	N	+	+	N	+	N	+
CoSy	N	+	+	N	+	+	N
Emergent							
AAR	Y	Y	Y	N	N	+	Y
Global Workspace	+	+	+	Y	N	Y	Y
I-C SDAL	+	+	+	+	+	Y	Y
SASE	Y	Y	Y	+	Y	Y	Y
Darwin	Y	Y	+	N	+	Y	Y
CogAff	Y	Y	Y	Y	Y	Y	Y
Hybrid							
HUMANOID	Y	Y	Y	+	+	+	N
Cerebus	Y	Y	Y	+	+	N	N
Cog	Y	Y	Y	+	N	N	N
Kismet	Y	Y	Y	N	N	Y	N
LIDA	+	+	+	Y	Y	+	+
CLARION	N	+	+	Y	Y	+	+
PACO-PLUS	Y	Y	Y	N	Y	N	N

The code used in Table 9.1 is the following: "Y" denotes that the feature is strongly addressed in the architecture, **"N"** denotes that the feature is not addressed at all, and "+" indicates that the feature is weakly addressed. The above Table verifies the fact that the available architectures have different capabilities. Many of these architectures are continuously enhanced and improved.

9.8 Concluding Remarks

Cognitive and intelligent control architectures are the cornerstone of mental robots, and their proper choice facilitates the design, implementation and evaluation of the robots. The architecture and its implementation are typically very strongly tied together in "built it and make it work" fashion. In this chapter we have provided an overview of principal architectural aspects of cognitive-intelligent robots, particularly: (i) the generic requirements and capabilities of cognitive robot architectures, (ii) a number of architecture taxonomies, (iii) some variants of meta-cognition in robot system architectures, (iv) a tour to the main conceptual and structural features of the Cog/Kismet behavior-based architecture, the SOAR and ICARUS symbolic (cognivist) architectures, the Global Workspace emergent architecture, and the LIDA hybrid (cognivist/emergent) architecture. The evaluation and validation architectural issues of cognitive architectures were also discussed along with a comparative review of twenty cognitive architectures.

The two major design properties that are required for the development of any cognitive architecture are *memory* and *learning*. A taxonomy of cognitive architectures on the basis of these features was given in Figure 9.4. More complete taxonomies may be obtained by adding more taxonomy criteria based on other features such as brain realization, adaptation, flexibility, evolution, and modularity. Thus possible directions for further research seem (among other) to include deeper studies and understanding of human memory and learning on the one hand, and integrate more architectural features for the study, design and evaluation of cognitive architectures on the other.

References

[1] P. Langley, J.E. Laird and S. Rogers, Cognitive architectures: Research issues and challenges, Cognitive Systems Research, 10, 2009, 141–160.
[2] J. Levesque and G. Lakemeyer, Cognitive robotics, Handbook of Knowledge Representation (Ch.24), Amsterdam: The Netherlands, 2006.

[3] G. Ernst and A. Newell, GPS: A Case Study in Generality and Problem Solving, New York, U.S.A.: Academic Press, 1969.

[4] R.C. Arkin, Behavior-Based Robotics, Cambridge, MA, U.S.A.: The MIT Press, 1998.

[5] A.A.D/Medeiros, A survey of control architectures for autonomous mobile robots, J. Brazilian Computer Soc., 4(3), 1968, 1–10.

[6] W. Duch, R.J. Oentaryo and M. Pasquer, Cognitive architectures: Where do we go from here? In: P. Wang, B. Goertzel and S. Franklin (eds.), Frontiers in Artificial Intelligence Application, Vol. 171, Amsterdam, The Netherlands: IOS Press, 2008, 122–136.

[7] G.N. Saridis, Toward the realization of intelligent controls, Proceedings of IEEE, 67(8), 1979, 1115–1133.

[8] G.N. Saridis, Foundations of intelligent controls, Proc. IEEE Workshop on Intelligent Control, Troy, N.Y., U.S.A., 1985, 23–28.

[9] G.N. Saridis, Analytical formulation of the principle of increasing precision and decreasing intelligence for intelligent machines, Automatica, 25(3), 1989, 461–467.

[10] A.M. Meystel, Autonomous Mobile Robots: Vehicles with Cognitive Control, Singapore: World Scientific, 1991.

[11] I. Ayari and A. Chatti, Reactive control using behavior modeling of a mobile robot, Intern. J. Computers, Communication and Control, 2(3), 2007, 217–228.

[12] R.A. Brooks, A robust layered control system for a mobile robot, IEEE J. Robotics and Automation, RA-2, 1986, 14–23.

[13] R.A. Brooks, Intelligence without reason, AI Memo, No.1293, AI Lab., MIT, 1991.

[14] R.C. Arkin, Motor Schema-based mobile robot navigation, Intern. J. Robotics Research, 8(4), 1989, 92–112.

[15] R.C. Arkin, Cooperation without communication: Multi-agent schema based robot navigation, J. Robotic Systems, 9(2), 1992, 351–364.

[16] A. Sloman, Varieties of meta-cognition in natural and artificial systems, In: M.T. Cox and A. Raja (eds.), Proc. AAAI 2008 Conf.: Workshop on Meta-reasoning, Menlo U.S.A.: AAAI Press, 2008, 12–20.

[17] A. Sloman, Architectural-based conceptions of mind, In: The Scope of Logic Methodology, and Philosophy of Science (Vol. II), Synthese Library, Vol. 316, Dordrecht: Kluwer, 2002, 403–427. http://www.cs.bham.ac.uk/research/projects/cogaff/00-02.html#57

[18] A. Sloman and B.S. Logan, Evolvable architectures for human-like minds, In: G. Hatano, N. Okada and H. Tanabe (eds.), Affective Minds, Amsterdam: Elsevier, 2000, 169–181.

[19] A. Sloman and M. Scheutz, A framework for comparing agent architectures, Proc. UK Workshop on Computational Intelligence, Birmingham, U.K., September, 2002.

[20] A. Sloman, The Computer Revolution in Philosophy, Sussex, U.K.: Harvester Press & Humanities Press, 1978.

[21] D. Vernon, G. Metta and G. Sandini, A Survey of artificial cognitive systems: Implications for the autonomous development of mental capabilities in computational agents, IEEE Trans. Evolutionary Computation, 11(2), 2007, 151–180.

[22] D. Vernon, C. von Hofsten and L. Fadiga, A roadmap for cognitive development in humanoid robots, COSMOS: Cognitive Systems Monographs, Vol.11, Berlin, Germany: Springer, 2011.

[23] R.A. Brooks, The Cog Project: Building a humanoid robot, In: C. Nehaniv (ed.), Computation for Metaphors Analogy and Agents, Berlin, Germany: Springer, 1998.

[24] C.L. Breazeal, Designing Sociable Robots, Cambridge, MA: The MIT Press, 2002.

[25] G. Givalt, R. Chatila and M. Vaisset, An integrated navigation and motion control system for autonomous multisensory mobile robots, Proc. First Intern. Symposium on Robotics Research, M. Brady and R. Paul (eds.), MIT, Cambridge, MA, U.S.A., 1983.

[26] R. Chatila, R. Alami and H. Haruelle, Integrated planning and execution control of autonomous robot actions, Proc. IEEE Intern. Conf. on Robotics and Automation (ICRA'92), Nice, France, 1992.

[27] J.E. Laird, P.S. Rosenbloom and A. Newell, SOAR: An architecture for general intelligence, Artificial Intelligence, 33, 1987, 1–64.

[28] J.L. Laird, P.S. Rosenbloom and A. Newell, Chunking in SOAR: The anatomy of general learning mechanism, Machine Learning, 1, 1986, 11–46.

[29] J.E. Laird, Extending the SOAR cognitive architecture, Proc. Artificial Intelligence Conf., Memphis, TN, U.S.A., (Amsterdam, The Netherlands, IOS Press), 2008.

[30] S. Nason and J.E. Laird, SOAR-RL: Integrating reinforcement learning with SOAR, Cognitive Systems Research, 6, 2005, 51–59.

[31] P. Langley, K. Cummings and D. Shapiro, Hierarchical skills and cognitive architectures, Proc. 26th Annual Conf. of the Cognitive Science Society, Chicago, IL, U.S.A., 2004, 779–784.

[32] P. Langley and D. Choi, Learning recursive control programs from problem solving, J. Machine Learning Research, 7, 2006, 493–518.

[33] M. Shanaham and B.J. Baars, Applying global workspace theory to the frame problem, Cognition, 98, 2005, 157–176.

[34] M. Shanaham, A cognitive architecture that combines internal simulation with a global workspace, Consciousness and Cognition, 15, 2006, 433–449.

[35] B.J. Baars, G. Edelman and J. Galley, Biology of consciousness, Frontiers in Psychology, 2(4), 2011.

[36] B.J. Baars, In the Theater of Consciousness: The Workspace of the Mind, Oxford, U.K.: Oxford University Press, 1997.

[37] S. Franklin, The LIDA architecture: Adding new modes of learning to an intelligent, autonomous, software agent, Proc. Intern. Conf. on Integrated Design and Process Technology: Society for Design and Process Science, San Diego, CA, U.S.A., 2006.

[38] U. Ramamurthy, B.J. Baars, S.K. D'Mello and S. Franklin, LIDA: A working model of cognition, In: D. Fum, F. DeMissier and A. Stocco (eds.), Proc. 7th Intern. Conf. on Cognitive Modeling, Trieste, Italy: Edizioni Goliardische 2006, 244–249.

[39] R. Brooks and L.A. Stein, Building brains for bodies, Autonomous Robotics, 1, 1994, 7–25.

[40] J. Snaider, R. McCall, S. Strain and S. Franklin, The LIDA Tutorial: Version 1.0, Cognitive Computing Research Group, The University of Memphis, TN, U.S.A.

[41] R.G. Simmons, Concurrent planning and execution for autonomous robots, IEEE Control Systems Magazine, 12(1), 1992, 46–50.

[42] J.H. Connell, SSS: A hybrid architecture applied to robot navigation, Proc. 1992 IEEE Conf. on Robotics and Automation (ICRA'92), 1992, 2719–2724.

[43] B.R. Duffy, M. Dragone and G.M.P. O'Hare, The social robot architecture: A framework for explicit social interaction, Proc. CogSci 2005 Workshop Android Science: Towards Social Mechanism, Stresa, Italy, 2005.

[44] D.E. Meyer and D.E. Kieras, A computational theory of executive cognitive processes and multiple task performance, Part I: Basic mechanisms, Psychological Review, 104(1), 1997, 3–65.

[45] J. Rosbe, R.S. Chong and D.E. Kieras, Modeling with perceptual and memory constraints: An EPIC-SOAR mode of a simplified 'en route' air traffic control task, SOAR Technology Inc. Report, Ann Arbor, Michigan, U.S.A., 2001.

[46] R. Alami, R. Chatila, A. Clodic and S. Fleury, Towards human-aware cognitive robots, Proc. 5th Intern. AAAI Spring Symposium: "To Bodily Go Where No Human-Robot Team Has Gone Before", March, 2006, 39–46.

[47] A. Sloman, Architectural requirements for human-like agents both natural and artificial: What sort of machines can love? In: K. Dautenhahn (ed.), Human Cognition and Social Agent Technology, Advances in Consciousness Research, Amsterdam, The Netherlands: John Benjamins, 2000, 163–195.

[48] C. Wei and K.V. Hindricks, An agent-based cognitive robot architecture, In: M. Dastani, J.F. Hübner and B. Logan (eds.), ProMAS 2012, Lecture Notes in Artificial Intelligence 7837, Berlin, Germany, Springer, 2013, 54–71.

[49] A. Nestor and B. Kokinov, Towards active vision in the DUAL cognitive architecture, Intern. J. Information Theories and Applications, 11, 2004, 9–15.

[50] P. Wang, Rigid Flexibility. The Logic of Intelligence, Berlin, Germany: Springer, 2006.

[51] J. Carbonell, C. Knoblock and S. Minton, PRODIGY: An integrated architecture for planning and learning, In: K. VanLehn (ed.), Architectures for Intelligence, Hillsdale, NJ, U.S.A.: Lawrence Erlbaum, 1990.

[52] S. Minton, Quantitative results concerning the utility of explanation-based learning, Artificial Intelligence, 42, 1990, 363–391.

[53] A. Oreback and H.L. Christensen, Evaluation of architectures for mobile robots, Autonomous Robots, 14, 2003, 33–49.

[54] E. Coste-Manière and R. Simmons, Architecture the backbone of robotic systems, Proc. 20th IEEE Intern. Conf. on Robotics and Automation (ICRA'2000), San Francisco, CA, U.S.A., April, 2000, 67–72.

[55] S. Limsoonthrakul, M.L. Dailey and M.A. Sirsupunduit, A modular system architecture for autonomous robots based on blackboard and publish-subscribe mechanisms, Proc. IEEE Intern. Conf. on Robotics and Biomimetics (ROBIO 2009), Bangkog, February, 2009, 633–638.

[56] S. Franklin, A consciousness artifact, J. Consciousness Studies, 10, 2003, 47–66.

[57] B.J. Baars and S. Franklin, An architectural model of conscious and unconscious brain functions: Global workspace theory and IDA, Neural Networks, 20, 2007, 955–961.

10

Additional Issues

"Consciousness" is a word worn smooth by a million tongues.
Depending upon the figure of speech chosen it is a state of being,
a substance, a process, a place, an epiphenomenon, an emergent
aspect matter, or the only true reality.
George Miller

"Consciousness" refers to those states of sentience and awareness
that typically begin when we awake from a dreamless sleep
and continue unitl we go to sleep again, or fall into a coma
or die or otherwise become "unconscious".
John Searle

The presence of mental images and their
use by an animal to regulate its behavior,
provides a pragmatic definition of "consciousness".
D.R. Griffin

10.1 Introduction

In this book we have examined the six major constituents of mental robotics (cognition, intelligence, autonomy, consciousness, conscience, ethics) and provided answers to fundamental ontological, epistemological, and ethical questions about them, Neighboring fields of research (philosophy, psychology, neuroscience, evolutionary biology, and other) offered new insight about how intelligence works in the real world. The new way of thinking in robotics and artificial intelligence has contributed substantially in the understanding of the nature of cognition, and the development of intelligent and cognitive robots by linking more closely sensing and acting. Two basic empirical questions about mental robots are: (i) whether robots and computers will be able to perform particular tasks that human performs, and (ii) if this can be done, whether robots and computers perform them in the same way that humans do. But the most important question is "whether a robot's/computer's ability to perform advanced human-like tasks, as a human would perform them, shows that the robot/computer has a mind, i.e., it thinks, is intelligent, is conscious, creative, and autonomous with free will. Other philosophical questions relevant to mental robots are:

- Does the body have an impact on how we think?
- Must a mental robot use a language?
- Is embodied cognition possible to be achieved in mental robots?

To answer the first question, psychologists test the influence of defined movements on memory *"embodied cognition"*, and neuroscientists investigate the role of the brain's motor system in cognitive skills (such as *language recall*).

In this chapter we present a number of conceptual issues complementary to those studied in the previous chapters. Specifically, the following topics are considered:

- Consciousness theories (HOT, GW, IIT, BC).
- Theory of mind and robot development.
- Robot attention and shared attention.
- Duty, legal, and ethical issues.
- Philosophical and ethical issues of cyborgs.

10.2 Consciousness Theories

The principal theories of consciousness are the following:

- Higher-order thought *(HOT)* theory [1–4].
- Global workspace theory [5, 6, 43, 44].
- Integrated information theory [7, 8].
- Biological theory *(BT)* [9–12].

Here, we will outline the main conceptual elements of them.

10.2.1 HOT: Higher-Order Thought Theory

According to this theory, an experience is *phenomenally conscious* only in virtue of another state which is about the experience. The higher-order state is considered to be a thought. Hence the name *higher-order thought* (**HOT**) is used. For example, a conscious experience of green consists of a representation of green in visual system which involves a thought for green to the effect that the person has the experience of green. Phenomenal consciousness refers to *"what is like"* to possess an experience. Always, when we talk about the physical basis of phenomenal consciousness (or, simply consciousness) we have to consider the so called *"explanatory gap"* (which refers to *"nothing that we know"*, i.e., nothing that we have been able to hypothesize or even fantasize). The problem of closing the explanatory gap (known as *"Hard Problem"* [13]) involves the following four issues:

- We do not see a hint on a solution.
- No good argument that there is no solution that another kind of being could grasp is available.
- The explanatory gap is not intrinsic to consciousness.
- To recognize the above three issues we need a special theory of consciousness.

Accepting that consciousness is in one or another sense based in the brain, the problem arises of why the brain basis of this experience is the basis of this experience rather than another one, or none. This obviously implies that nothing now known gives a hint of an explanation.

According to HOT theory, consciousness of say, blue is a matter of three items:

- A higher-order thought.
- A representation with the content blue.
- An aboutness relation between the above two items.

These ingredients may exist individually without any consciousness. If a person has an unconscious representation of blue, and then creates an unconscious thought about the representation of blue, the representation of blue automatically becomes conscious. This means that putting together ingredients not conscious in themselves (thought, aboutness, representation), automatically creates consciousness. But HOT theory cannot assure that neuroscience will find some magic flash that appears when those items combine.

10.2.2 Global Workspace Theory

In the *global workspace* (**GW**) model, consciousness is regarded to be a place where unconscious elements interact with the system in order to access the *global workspace* [44]. This is based on Baddeley and Hitch's *working memory* (**WM**) concept which has considerably influenced the development of *"integrative models of cognition"* (Figure 10.1). Its basic components are the *phonological buffer* (which can be tested by silent rehearsal of words or

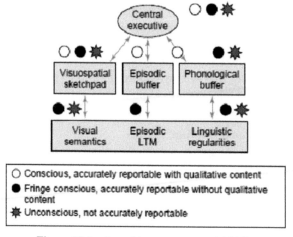

Figure 10.1   A working memory model [44].

Source: http://core.ac.uk/download/pdf/86955.pdf

numbers), the *visuospatial sketchpatch* (tested through mental images in problem solving), and the *central executive* (verified by voluntary manipulations of WM functions).

A simple representation of the global workspace model is shown in Figure 10.2, where the unconscious specialized processors constitute the basic components in an entity. These processors work in parallel with other

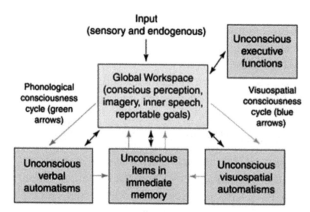

Figure 10.2   Global workspace model. Conscious elements play an essential role in the distributed working memory. Thick arrows represent phonological and thin arrows represent visuospatial consciousness cycles [44].

Source: http://www.neurdon.com/wp-content/uploads/2009/07/baars-franklin_wm2003diagram.jpg

processors in order to achieve more complex tasks. Consciousness is seen as a global workspace in the brain whose role is to focal contents that are widely distributed *("broadcast")* in many unconscious specialized components. The processors that fit these needs form a coalition. A specific coalition formed several times is fixed into a *"context"*.

Thus a context involves several unconscious processors pursuing a common goal. Because a context is learned by consciousness, while forming coalitions, it is unconscious. This means that *"unconscious networks"* (contexts) shape conscious contents, i.e., contexts work together jointly to constraint conscious events. Actually, motives and emotions can be regarded as goal contexts, and executive functions work as hierarchies of goal contexts. A hierarchy is the organization of several contexts that can be active at the same time.

In Figure 10.3 it is demonstrated how a global workspace capability associated with consciousness could recruit standard memory functions (e.g., rehearsal or visual imagery).

These functions are considered *"slave systems"* in working memory, operating as semi-autonomous distributed networks. The architectural aspect of the global workspace model can be implemented using neural concepts, and also

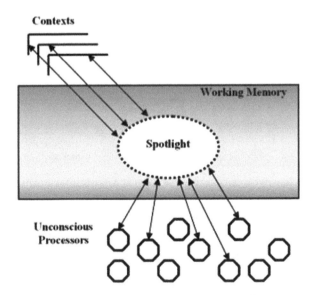

Figure 10.3 Baar's global workspace theory diagram.

Source: http://www.conscious-robots.com/images/stories/logos/gwt_en.jpg

in electronic computers. For this reason philosophers call it *"functionalism"* according to which consciousness is characterized by an abstract construct not needing the details of the neuroscience [12]. A neuronal implementation of a global workspace, the so called *"neuronal global workspace"* model is shown in Figure 10.4 [14]. In this model, sensory stimuli mobilize excitatory neurons with long-range cortico-cortical axons, that produce global activity patterns among workspace neurons.

The global neuronal workspace model is a *meta neural network*, i.e., a network of neural networks composed of a very large number of integrate- and fire neurons, and forecasts that conscious presence is a nonlinear function of stimulus salience (i.e., a gradual increase in stimulus visibility should be accompanied by a sudden transition of the neuronal workspace into a corresponding activity pattern).

10.2.3 Integrated Information Theory

The *information integrated theory of consciousness* (**IITC**) claims that consciousness corresponds to the capacity of a system to integrate information. The information integration capability of a system can only be acquired when a large repertoire of states exists, with states of each element being causally dependent on the states of other elements. The IITC proposes a new measure

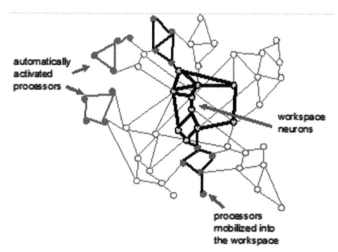

Figure 10.4   The neuronal workspace model.

Source: http://www.scholarpedia.org/article/Models_of_consciousness#Global_workspace_models
http://www.unicog.org/publications/DehaeneChangeux_ReviewConsciousness_Neuron2011.pdf

for the *"quantity"* of consciousness generated by a system. This measure M is defined as the amount of causally effective information that can be integrated across the weakest link of a system. The integrated information theory of consciousness is based on the following postulates:

- Consciousness exists
- Conscious states are highly differentiated (i.e., they are informationally very rich)
- Consciousness is structured
- Consciousness is informative
- Consciousness is integrated, i.e., whatever information a human is conscious of is entirely and completely presented to his/her mind (it cannot be subdivided)

IITC embraces the theoretical sense of information, i.e., information is the reduction in uncertainty regarding the state of a variable, and conversely it is what increases in specifying a variable with a growing number of possible states. In a system that consists of connected nodes which contain information and causally influence other nodes (called *"mechanisms"*), the information among them is said to be integrated if there is a greater amount of information in the *"reprertoire"* of an entire system regarding its previous state than there is in the sum of all the mechanisms considered individually.

10.2.4 Biological Theory

According to the *biological theory* (**BT**) of consciousness, consciousness is some kind of biological state of the brain. Its roots go back to *Democritus*, and it was set in modern form by Place (1956, [9]), Smart (1959, [10]) and Feigl (1958, [11]). BT is also discussed in [12, 13]. This theory, also known as *identity theory,* came into existence as a reaction to *behaviorism*. Behaviorism does not accept the existence of mind, emphasizing only on behavior. This is in contrast to the dualistic point of view of the *mind-body* problem first formulated by *Rene Descartes*. According to Place, *behaviorism* and *dualism* could not give a satisfactory solution of the consciousness problem. So, with his *identity theory* says that mental states and processes, and the brain states and processes are identical. Smart did not accept Place's theory to the full extent, and claims that his arguments for identity theory are very much different from Place's arguments [10].

To justify Place's theory and make it unobjectionable Smart has addressed a number of objections to the theory. Three of them are the following [18]:

1. *An illiterate man may not know anything about neuropsychology but he can perfectly talk about after-images or how things look of feel to him. Therefore, when someone describes his sensations, the things he is talking about cannot be processes in the brain.*

Smart states that a related statement of the form *"A is identical with B"* can be there and that "something is an A" may be known by a person without knowing that it is also B. For example, someone can talk about lighting without any knowledge about electricity.

2. *To report a sensation is not a report of brain process.*

Smart says that this objection means that having an *after-image* does not mean that someone has such and such brain process. But the fact is that what one reports is a brain process. For example, the meaning of the *"Evening Star"* and the meaning of the *"Morning Star"* do not mean the same thing but the two stars are one and the same thing. Thus according to Smart, the statement "the Evening Star and the Morning Star are one and the same thing is" is a contingent one :

3. *The brain process occurs in physical space but the after-image is not, and therefore is not correct to say that after-image is a brain process.*

Smart says that his argument is not like that the after-image is a brain process but rather that experience of having an after-image is a brain process. Our introspective report is about *"experience"*. Smart also claims that his thesis is essentially a supplement of the thesis: "The *Mental* is the *Physical*" of Feigl [11]. Actually, Smart followed Place in applying the "Identity Theory" only to those mental concepts considered resistant to behaviorist treatment (mainly sensations).

10.2.5 Comparison Remarks

Here, we will give a few comparison remarks on the above theories of consciousness [16].

- From the Biological Theory point of view the Global Workspace and HOT theories are what consciousness *does* rather than what consciousness *is*, and these theories leave out too many details of the actual working of the brain to be adequate theories of consciousness.
- From the HOT point of view, the biological and global workspace accounts underestimate the role of cognition in consciousness.
- By many scientists the global workspace and HOT accounts are viewed as superior to the biological theory in that the biological theory allows

for the possibility that a person could have a phenomenally conscious state that the person does not know about.

- From the biological and global workspace views, HOT sees consciousness as more intelligent than is.
- The HOT and global workspace theories link consciousness to the ability to report it more tightly than does the biological theory.
- In HOT account, reporting is just expressing the higher-order-thought that makes the state conscious, so the underlying basis of the ability to report comes with consciousness itself.

Further discussions on Place's thesis that consciousness is a brain process are provided in [17, 18].

10.3 Theory of Mind and Robot Development

10.3.1 Theory of Mind

The *theory of mind* (or, in full, *theory of other minds*) is one of the fundamental skills of humans. This theory is an ingredient of human social dynamics, and studies the human ability to correctly attribute percepts, goals and beliefs to other persons. It allows us to understand the expressions and actions of others within an intentional or goal-directed framework (the so-called *intentional stance*) [19]. Recognizing that other persons have knowledge, perception, and intentions that differ from our own is a crucial step in children development (self-recognition, linguistic acquisition, imagination, and creative play). Emerging experimental efforts in cognition simulation and simulated social integration provide good opportunities to study the *theory of mind* (**ToM**) in interactive, *"first-person"* and *"second person"* scenarios while affording greater experimental control.

A good working definition of ToM can be derived by answering the question *"How do humans accurately infer the mental states of others?"* From this consideration, the following components of interaction have emerged as traces to ToM [10]:

- *Knowledge of the shared context:* ToM is a manifestation of cognition and is situated in the context of the surrounding world. For example, during a conversation people must quickly infer the thoughts, beliefs, goals and emotions in order to formulate an appropriate response, something that requires the integration of cues from the conversational partner, and the knowledge about the relationship between individuals.

- *Perception of social cues:* Although mental states are inherently cognitive phenomena, humans have a sophisticated range of behaviors such as gaze cues, vocal cues, and facial expressions through which they express their mental states. Gaze perception and, emotion recognition have received much attention in the field, and actually there exist several tasks to study both gaze behavior and emotion recognition.
- *Interpretation of actions:* Several tasks are available to evaluate human abilities to infer mental states by passive observation of behavior. Tasks requiring the interpretation of actions are profitably used in developmental literature, but studies of joint (shared) action can provide additional insight into how humans interpret actions to infer the mental states of others.

10.3.2 Leslie's and Baron-Cohen's Models of Theory of Mind

Two well- known models of ToM have been developed by Leslie and Baron-Cohen [22–24]. A brief exposition of them is as follows:

- **Leslie's model of ToM**   According to Leslie, the world is naturally decomposed into three classes of stimuli based upon their causal structure, namely:

Mechanical agency: The theory of body module is concerned with events that can be best described by mechanical agency, i.e., they can be explained by the principles and rules of *mechanics.*

Actional agency: The second module of ToM involves the actions of the agents, i.e., events are explained in terms of the intent and goals of agents (their actions).

Attitudinal agency: Events are explained in terms of the *attitudes* and beliefs of agents.

- **Baron-Cohen's model of ToM**   This model assumes that two forms of perceptual information are available as input, namely [24]:
1. *Stimuli in the visual, auditory, and tactile perceptual spheres that have self-driven motion.*
2. *Visual stimuli that have eye-like shapes.*

According to Baron-Cohen, the set of precursors to a ToM, which he calls the *"mindreading system"* can be decomposed in the following four distinct modules:

1. *Intentionality detector* (**ID**): This module interprets self-driven motion of stimuli through the primitive volitional mental states of goal and desire.

2. *Eye direction detector* **(EDD)**: This module detects the presence of eye-like stimuli in the visual field and computes whether the eyes of an individual are looking at the partner's face, and spend more time gazing at the eyes that at other parts of the face.
3. *Shared attention mechanism* **(SAM)**: This module takes the dyadic representations of ID and EDD and produces triadic representations like "George sees (I see the girl)".
4. *Epistemic mental states representation:* In this module a way for representing epistemic mental states in other agents is provided, along with a mechanism for tying together our knowledge of mental states into coherent whole as a usable theory.

Baron-Cohen argues that both ID and the basic functions of EDD are available to infants in the first nine months of life, SAM develops between 9 and 18 months, and the fourth capacity develops from 18 months to 48 months. This ToM model was also proved very useful in the development of social skills in autistic individuals (allowing a comparison with normal individuals).

10.3.3 Developmental Robots

Robot mental development is based on the theory of mind. When robots can autonomously develop capabilities such as speech, vision and language, humans can train them using their own communication modes. Traditionally, a robot is not autonomous when it develops its skills and capabilities, but a human is autonomous over its entire lifelong mental development. Thus, computational modeling of human neural and cognitive development helps decisively in robot development. To be successful, cognitive psychology methods are needed in order to advance from explaining psychological phenomena in specific controlled settings toward deriving underlying computational principles of mental development which are applicable to general settings.

The basic steps for constructing developmental robots are [25]:

- Design a body according to the robot's ecological conditions on land, water, or air.
- Design a developmental program.
- Running of the developmental programs starts at the robot's birth.
- To develop the robot's mind, humans mentally "raise" the developmental robot through real-time interaction.

The above procedure suggests that robots should be designed to go through a long period of autonomous mental development from "infancy" to "adulthood", in order to become able to autonomously "live" in their environment and become intelligent and smart on their own, with some supervision by humans. Like traditional programs, developmental robot programs should be body-specific, or specific to robot species. Developmental robot programs must have, among others, the following features:

- No task specific
- Tasks are unknown
- Ability to represent unknown task
- Animal-like online learning
- Open-ended learning

These capabilities are not required in traditional robot programs. Here, it is noted that a robot cannot learn complex skills successfully without first learning necessary simpler skills. For example a robot can learn how to write, if first learns how to hold a pen.

Basic skills for robot development include:

- Attention direction
- Action suggestion
- Positive/negative feedback

The issue of *attention* will be studied in the next section.

Among others, developmental robotics differs from traditional robotics and artificial intelligence in the following aspects [26].

- The body structure and environment play a dominant role as causal elements in the emergence of organized behavior and cognition, and requires their explicit inclusion models of emergence and development.
- The artificial cognitive systems of the robot are realized by initiating and maintaining a developmental process during which the robot interacts with its *physical environment* (through its body, suitable tools, or other artifacts) and with its *social environment* (i.e., with people, other robots, or simulated agents).

Cognition, after all, is the outcome of a self-organization (spontaneous emergence of order) and co-development between a developing organism and its surrounding environment. This process is known as *"cognitive incrementalism"*, and designates the process of starting a minimal set of functions and building increasingly more functionality in a step-by-step way on top of structures that already exist in the system.

Developmental robotics involves the following (possibly overlapping) fields:

- *Socially oriented interaction* Learning and development is achieved via social interaction with humans or other robots.
- *Non-social interaction* Learning and communication are achieved via the strong coupling between sensor and motor processes and local (inanimate) environment, not involving any interaction with humans or other robots.
- *Agent-centered sensorimotor control* Here, the body capabilities, the changes of morphology (e.g., strength of effectors), and their effects on motor skill acquisition, supervised learning schemes, etc., are studied. Skills like self-exploration, categorization of motor patterns, learning to walk etc. are acquired.

Two questions referred to developmental robotics are:

1. Could there be laws governing developmental robots?
2. Could there be a theory?

So far, no complete answers to these questions were given. An obvious way to develop such theory is to point out a set of candidate design principles, abstracted from biological systems (observation of human and animal development). Theories of this kind are well-known as *"biologically inspired theories"*. Further success in developmental robotics will depend on the extent to which theorists and experimentalists will be able to formulate universal principles covering the entire range of levels on which developmental robots operate.

10.4 Robot Attention

10.4.1 General Issues

Attention is the process by which a being or agent allocates perceptual resources to explore a subset of the surrounding world in detriment of others [27, 28]. Attention is distinguished in:

- **Overt attention** The allocation of perceptual resources consists of an explicit reorientation of sensors (visual, auditory, or head reorientation) to a specific part of the environment.
- **Covert attention** The deployment of computational resources for processing a specific part of the stream of sensory information.

Thus, overt attention is the direct cause of a process called *active perception* [29]. Active perception can be either voluntarily or involuntarily driven, where in the second case active perception is characterized by the automatic orientation of sensors. In visual attention this is implemented using *gaze shifts* via head reorientations and rapid eye movements, and *fixations*. Typically overt processes follow covert processes in the overall timeframe of attention.

Attention is triggered in two different ways: namely (Figure 10.5):

- *Stimulus driven* (i.e., affected by bottom-up influences)
- *Goal directed* (i.e., affected by top-down influences).

Attention (i.e., the ability to "concentrate") is important to learning, since learning is optimally efficient when an individual is paying attention. The attention ability of selecting the potentially relevant parts out of a large amount of sensory data enables interaction with other persons via *"sharing attention"* to each other. This ability is of great importance in robotics, where the computational modeling of human attention is a key issue in human-robot interaction. Actually, overt attention and active perception have a strong significance for social beings, and are the *"precursor of intent"*. Overt attention is, in fact, intentionally directed perception (called also *intentional*

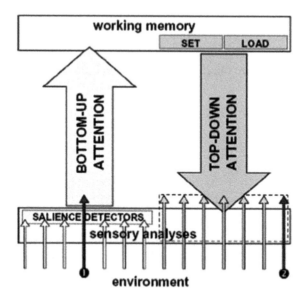

Figure 10.5   Illustration of bottom-up and top-down attention processes. Bottom-up attention is triggered by sensory stimuli themselves. Top-down attention is driven by cognitive goals activated in working memory.

Source: https://healthskills.files.wordpress.com/2009/05/attention-processing.jpg?w=640

stance or *intentional contingency*). The *center of spotlight* of attention (overt or covert) is known as the *"focus of attention"* (**FOA**). In life only the FOA is analyzed in detail. Regions other than the FOA are mostly ignored at any specific time instant, producing the so-called *"change blindness"* for peripheral stimulation.

10.4.2 Shared Attention

Shared or joint attention is the cornerstone of human cognition social intelligence and interaction. Shared attention is defined as orienting or re-orienting attention to a target because it is the object of another person's attention. A basic manifestation of shared attention is *gaze following* which is the result of adaptive compromise between human's perceptual limitations and our social tendencies. Gaze following emerges around nine months after infants start scanning the internal contours of faces, and respond to caregivers breaking eye contact.

Modern mental robots detect and track people (a precursor of shared attention) using visual features (e.g., face templates, flesh color) similar to the features, not necessarily the same, used by infants. Infants prefer to respond to dynamic social events (e.g., head movement) and to sequential contingencies among social events.

A practical way for developing shared attention capabilities for robots is to use as a basis the shared attention skills of infants, namely [30]:

- **Contingency learning and intermodal synchrony** Young infants respond to the social contingencies or "games" established by caregivers. Contingency behavior appears to be a pervasive cue for shared attention.
- **Learning and affect** Reinforcement learning and statistical (Bayesian) learning are convenient typical learning modes for shared attention abilities development. Bayesian learning processes select the most predictive input pattern. This may explain why infants finally respond more reliably to pointing. Reinforcement contributes to improvement in gaze following, since infants and caregivers find the same kinds of events and objects reinforcing or interesting. Affect can suppress shared attention. For example, infants that are upset when parents broke eye contact, in order to look at a target, seldom followed gaze. Instead, they looked directly to the parent protesting orally in order to regain the parent's attention.
- **"Setting Conditions" and the "Game"** The context of infant caregiver interactions constitutes a key issue in shared attention. This includes

the temporal context of social events before the shared attention, the spatial setting of infant-caregiver-distal target, extraneous goals and activities, and the occurrence of environmental distractions. A *"game"*, i.e., setting condition, is particularly important and consists of the earliest face-to-face infant-caregiver interactions that are carried along attention alteration between an individual and distal object-event. A game of this type can be tested via robotic simulations, by varying the information given in the game.

Three questions that must also be answered in infant development via shared attention are the following:

- **What must infants represent to share attention?**

Some theorists have argued that infants have innate abstract social representations. Also, there is evidence that infants cannot reliably understand the relation between seeing and gaze until late in the second year, and, in fact, there is no evidence of an abstract theory of mind prior to about 18 months.

- **Does "the game" provide sufficient information for shared attention to develop?**

In about 10–12 months, infants shift from interacting with either people or objects, to a coordinated interaction between objects and people. But it is still unclear how this develops.

- **What are the "setting conditions" for shared attention?**

In other words, what impact do various social interactions have on infants emerging knowledge of caregivers, self, and effects on others? How does the infant's posture mode of carrying or transportation, and care schedule affect the emergence of shared attention?

Figures 10.6–10.8 show three instances of robot-human shared attention.

10.4.3 An Example: Conversational Robot

In [31, 32], a conversational robot was developed using a set of attention gestures (called *"engagement behaviors"*), in people-robot dialogs, that provide critical cues to the FOA of the participants (Figure 10.9). Engagement behaviors are those behaviors by which interlocutors start, maintain and end their perceived connection to one another. The process by which conversational participants exploit to perform this connection is the process of *"engagement"*.

The robot interacts collaboratively with a single user through the following:

Figure 10.6 Robot-human shared attention. MIT's Leo social robot is looking at the button pointed by a human.

Source: http://www.slashgear.com/gallery/data_files/7/4/MIT_Leonardo_robot_2.jpg

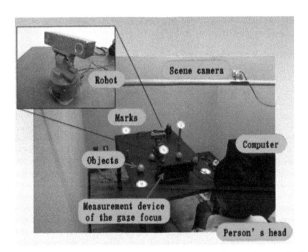

Figure 10.7 A second instance of robot-human shared attention. A human looks at four static balls on the table. The robot gazes at the human and the balls alternatively [45].

Source: http://siva.w3.kanazawa-u.ac.jp/image/lisboa01.pdf

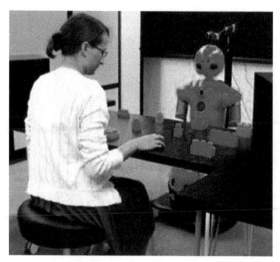

Figure 10.8   The Wakamaru robot uses gaze cues to establish shared attention in a sorting task [46].

Source: www.ncbi.nlm.nih.gov/pmc/articles/PMC3737477/#B46

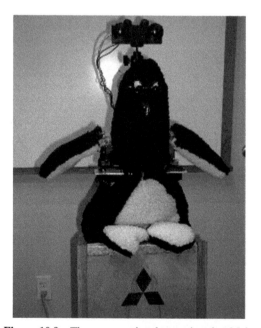

Figure 10.9   The conversational penguin robot Mel.

Source: http://web.cs.wpi.edu/~sidner/6-sidnerspring07final.pdf

- Spoken language: Understanding and generation using a mouth that opens and closes in coordination with the robot's expressions.
- Gestures with their appendages.
- Head gestures to track the user and to turn to look at objects of interest in the interaction.
- Recognition of user head gestures in looking at objects.
- Recognition of user head nods.

The robot is also able to initiate interactions with users, and performs typical pre-closings and goodbyes to end the conversation.

The architecture of the robotic system (Figure 10.10) involves two subsystems, namely: *conversational subsystem* and *sensorimotor subsystem*. The first subsystem employs a conversation and collaboration manager. The sensorimotor subsystem controls the physical hardware of the robot, also performing sensor fusion.

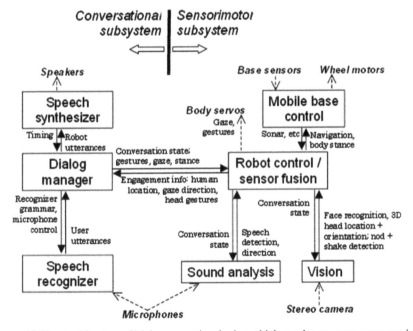

Figure 10.10 Architecture of Mel conversational robot which can demonstrate a communicative interaction demo with the user. Using a vision system, it can follow the looking gestures of a human, and the head mood of partners wishing to participate in the conversation.

Source: http://www.researchgate.net/publication/228031734_Attentional_Gestures_in_Dialogues_Bet ween_People_and_Robots

http://web.cs.wpi.edu/~sidner/6-sidnerspring07final.pdf

Mel can demonstrate a communicative interaction demo with the user. The robot is able to follow the looking gestures of a human using a vision subsystem. Participants indicate wish to engage in the conversation via head mooding. The robot is also able, via visual perception, to recognize nonverbal cues. The visual recognition of user gestures can be strongly improved via the dialog with the embodied conversation agent (Figure 10.11). Detailed results are provided in [31, 32].

Other examples of mental robots with various levels of cognition, intelligence, autonomy, consciousness and conscience are provided in [33–36].

10.5 Further Philosophical and Legal Issues of Mental Robots

10.5.1 General Aspects

Mental robots (social, medical, companion robots) raise philosophical, ethical and legal issues for individual citizens and entire societies. In particular, robot duties and rights, which emerge from autonomous decision and action capabilities lie at the center of the philosophical and legal concerns. Inherently, autonomy is a multi-scale process that depends on the layer of the control

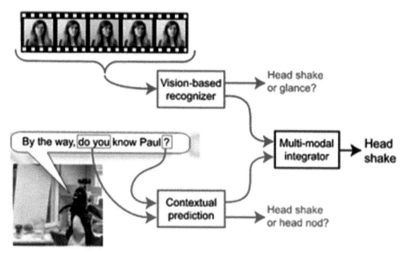

Figure 10.11   Contextual recognition of head gestures during face-to-face interaction with Mel. The contextual information from Mel's spoken utterances assists to disambiguate the listener's visual gesture.

Source: http://web.cs.wpi.edu/~sidner/morency-fourth-draft.pdf

hierarchy being awarded with a degree of autonomy, or involving environment or human influence in the decision loop. As we saw in Section 6.4, the levels of robot autonomy range from full autonomy (there is no human influence), to full non-autonomy (i.e., fully manual, human intervention). Between these two extremes there are several levels of supervisory, shared and human-assisted control (as shown in Figure 6.4).

Mental robots have the ability to develop and learn automatic behaviors and policies, and a higher degree of autonomy is usually accompanied by higher-order functions such as outdoor services, social behavior, and novel and emerging autonomous learning approaches.

Therefore, the investigation of *ethical legal* and *societal* (**ELS**) implications, particularly, with respect to liability issues, given the potentially unpredictable and unmanageable variety of learning experiences and operational situations of robots in unstructured environments, is of great importance for modern society.

These issues of mental robot autonomy include the following [37]:

- Duty issues
- Legal issues (legal responsibility)
- Ethical issues (ethical responsibility)

10.5.2 Duty Issues

A crucial point in autonomous intelligent robots concerns the duties that the robots can perform. The philosophical question here is the following: "Is it possible to assign duties to robots and open the discourse about this, without asking whether robots can support or not the set of conditions required (power-to-do, first-level-capability to perform precisely the action the robot is obliged to do, second-level capacity, i.e., ability to perform more than a single task)? This question is applicable to both external conditions and to internal conditions.

10.5.3 Legal Issues

The philosophical question here is whether one day robots will have rights, i.e., whether and how robots could become legal beings, rather than remaining an object under law. Another question is "whether a legal subjectivity (responsibility) for robots is needed or it is useful.

In Europe and the West civil liability in the manufacture and use of robots at present is classified as: contractual liability and non-contractual liability.

Contractual liability regards the robot as the object (the product) of the contract for sale between the seller (manufacturer) and the buyer (user). In other words, the robot is considered to be a *"consumer good, product or commodity"*, and the standard liability rules are applicable without any difficulty. Contractual liability occurs when the robot performance is not as agreed in the contract, even if the robot does not make a damage or harm.

Non-contractual liability occurs when the robot's action causes a "legally recognized damage" (e.g., infringement of a human right) to a human regardless of the existence or not of a contract. Here we distinguish two cases: (i) damage caused by a defective robot, (ii) damage caused by action or reaction of the robot with the human.

Further concerns arise from the possibility of assigning the legal personality (responsibility) to beings that are not humans including robots, computers and other systems. The basic question here is: "Can we speak about robots (computers or software programs) as a legal persons in legislation, since they are not people in real life? Many scientists and philosophers argue that the concepts of "person" and "human being" do not overlap at all. However the developments in science and biology and medicine have led to rethink, particularly at a philosophical level, these concepts and the opportunity of including some stages of human life in the category of *"person"*. Moreover, philosophers and scientists started to assume that other living beings (animals, plants) or even intelligent things (such as mental robots) would be considered as a person (see, e.g., [38]).

10.5.4 Ethical Issues

Robot ethics has been extensively studied over the years, and a plethora of theoretical and philosophical results are already available. However, robot ethics has not yet reached a practical level in Western society, although in the East (Korea and Japan) robot ethics is already applied to everyday human activity. In the West it is anticipated that autonomous mental robots may be regarded as legal *persons* in analogy to companies or corporations, and so they may enter a public register (similar to the commercial register).

The Korean *Robot Ethics Charta*, provides a set of principles about the human/robot ethics that assure the co-prosperous symbiosis of humans and robots. These principles are [39]:

- **Robot ethics principle**   The robot should obey the human as a friend, helper, and partner, and should no harm human beings.

- **User ethics principle**   The robot user must regard robot as his/her friend and forbid any illegal re-assembly or misappropriation of the robot.
- **Common human-robot ethics principle**   The human being and the robot are both deserved of dignity, information, and engineering ethics of life.
- **Manufacturer ethics principle**   Robots should be manufactured such that to defend human being dignity. Manufacturers are responsible for robot recycling and providing information about robot protection.
- **Government ethics principle**   The government and local authorities must enforce the effective management of robot ethics over the manufacturing and usage cycle.

Clearly, the above Korean human-robot ethical principles have an emotional and social character rather than a legal one. An analogous robot ethics charta, compatible with current legislation, is perhaps required in the West.

10.5.5 Philosophical and Ethical Issues of Cyborgs

Critical philosophical and ethical questions are related to the development and realization of **cyborgs** (**cyb**ernetic **org**anisms) which range from the wearable exoskeleton to the prototype that reproduces the artificial sense of touch and vision. The *cyborg* field belongs to *bio-robotics*, which today has reached the possibility to allow the modifications of one's body following his/her desires. This has in a sense the result of a loss of human nature. Cyborgs may have abilities that exceed the figure of conventional robots (industrial, service, social). This is done by combining the artificial reality with the human and the natural one, so that the technique is embodied in the human and becomes an extension of the body (that refers to the concept of *bionic man*).

The cyborg technology has been made possible by the fact that in the brain central nervous system, bioelectrical signals can be connected directly to computers and robot parts that are either outside the body or implanted in the body. When one succeeds in decoding the brain's movement signals, then these signals can be directly connected to external electronic equipment (mobile phones, TV sets, etc.), such that it may be possible to control electronic devices with the power of thought alone, i.e., without the use of articulated language or external motor device. In other words, cyborg-type prostheses can also be *virtual*. In shaping the body a person has, he (she) can model it depending on what he (she) wants to be like so that in such paradigm it becomes what he (she) desires to appear like. As a result, the *being* becomes *mere appearance*, losing all the ways of true determination and proper speaking; man is not a being but a *"becoming"* in perceptual change, since he can become all and recreate himself as he likes to. Thus, in modern world "to **have**" is replaced by "to **become**", and so man loses his essence and remains only with the appearance.

In medicine, there are two principal types of cyborgs namely the *restorative* type and the *enhanced* type. Restoring technologies "restore lost function, organs, and limbs" such as to revert to a healthy or average level of function. The enhanced cyborg technologies follow the principle of *optimal perfor-mance*, i.e., maximizing output (the modification achieved) and minimizing input (the energy expended in the process). Cyborg technology is a two-sided word, i.e., it can be used for the good or for the bad.

The primary ethical concerns surrounding the development of cyborgs are focused on human dignity, human relations, protection of physical/bodily harm, and the management of the health and other personal data evaluation. In all cases the primary rules of roboethics should be respected (autonomy, non-maleficence, beneficence, justice, truthfulness, dignity).

In particular:

- Every person in the society should be free to decide the ways he/she wishes to exploit the opportunities offered by cyborg technologies, as much as this decision does not damage the life development or private lives of others. Of course, the abilities that nature gives to each person have differing conditions for pursuing life goals and desires.
- Cyborg technology implanted in or integrated with people should be used exclusively for the purpose of healing disease or remedying disability (e.g., for replacing naturally existing functions-sight, limbs, etc., for people that have lacked them from birth, or lost them due to various reasons).
- Cyborg technology should not give rise of black market. Private hospitals that offer cyborg technological interventions should satisfy the same restrictions, and the same valid purposes that apply in the public health system.

The main advantages of mixing organs with mechanical parts that are for the human health. Specifically:

- Persons that were subject to surgery for replacing parts of their body (e.g., hip replacement, elbows, knees, wrists, arteries, veins, heart valves) can now be classified as cyborgs.
- There are also brain implants based on the neuromorphic model of the brain and the nervous system. For example, there are brain implants that

help reverse the most devastating symptoms of the Parkinson Disease. A deaf person can have his inner ear replaced and be able to engage in telephone conversation (or hear music in future).

The disadvantages of cybernetic organisms include the following:

- Robots can sense the world in ways that human cannot (ultraviolet, x-rays, infrared, ultrasonic perception). Thus there is more dependence on cyborg technology.
- Intelligent robots can outperform humans in aspects of memory and mathematical/logic processing.
- Cyborgs do not heal body damage normally, but, instead, body parts are repaired. Replacing broken limbs and damaged armor plating can be expensive and time consuming.
- Cyborgs can think the surrounding world in multiple dimensions, whereas human beings are more restricted in that sense.

A thorough philosophical discussion about cyborgs and the relationship between body and machine is provided in [40], and a general scientific discussion about cyborgs and the future mankind is given in [41].

Two real-life cyborg examples are the following. More examples are provided in [42].

Example 1
Jesse Sullivan is one of the world's first cyborgs equipped with a bionic limb connected through a nerve-muscle graft. Sullivan is able to control his new limb with his mind, and also to feel hot, cold, and the level of pressure his grip is applying (Figure 10.12).

Example 2
Jens Naumann has lost his sight (in both eyes) due to a couple of serious accidents. He became the first person in the world to receive an artificial vision system, equipped with an electronic eye connected directly to his visual cortex through brain implants (Figure 10.13).

10.6 Concluding Remarks

In this chapter we have discussed a number of additional issues concerning mental robots in order to complement the material presented in chapters 4 through 9. Specifically, the issues considered are the following: (i) consciousness theories (higher-order thought, global workspace, integrated information theory, biological theory), (ii) theory of mind, (iii) developmental robotics,

Figure 10.12 A cyborg limb that can be controlled by a person's mind [42].

Source: http://www.mnn.com/leaderboard/stories/7-real-life-human-cyborgs

Figure 10.13 Jens Naumann sees with a cyborg (electronic) eye connected directly to his visual cortex [42].

Source: http://www.mnn.com/leaderboard/stories/7-real-life-human-cyborgs

(iv) robot shared attention, (v) duty, legislation, and ethical issues, and (vi) philosophical and ethical issues of cybernetic organisms (cyborgs). A few remarks for further research in the topics of the chapter are the following:

- The main requirement towards building conscious robots is to define consciousness as a computational construct as much accurately as possible. To do this strong interdisciplinary efforts are needed in order to *"mix"* functional approximations of cognition such as fuzzy logic, data fusion, parallel computing, and neural network in order to develop more powerful cognitive architectures.

- Although passive and reflexive methods used for studying the "theory-of-mind" have contributed substantially in the understanding of the mechanisms of inferring mental states, on-line and interactive methods to study ToM and social cognition are expected to provide enhanced and more powerful methods.
- To advance further the field of developmental robotics, methods of semiotic dynamics (continuous refining the meaning of symbols), creativity, curiosity, self-exploration, motivations, prediction, and action selection are required.

References

[1] A. Byrne, Some like it HO: Consciousness and higher-order thoughts, Philosophical Studies, 86, 1997, 103–129.

[2] A. Byrne, Review of Carruthers' phenomenal consciousness, Mind, 110, 2001, 1057–1062.

[3] W.G. Lycan, Consciousness and Experience, Cambridge, MA: MIT Press, 1996.

[4] P. Carruthers, Phenomenal Consciousness: A Naturalistic Theory, Cambridge, UK: Cambridge University Press, 2000.

[5] B. Baars, A Cognitive Theory of Consciousness, Cambridge, U.K.: Cambridge University Press, 1988.

[6] S. Dehaene, J.-P., Changeux, I. Nacchanche, J. Sackur and C. Sergent, Conscious, preconscious, and subliminal processing: A testable taxonomy, Trends in Cognitive Science, 10, 2006, 204–211.

[7] G. Tononi and G.M. Edelman, Consciousness and Complexity, Science, 282 (5395), 1998, 1846–1851.

[8] G. Tononi, Information Measures for conscious experience, Archives Haliennes de Biologie, 139(4), 2001, 367–371.

[9] U.T. Place, Is consciousness a brain process? British Journal of Psychology, 47, 1956, 44–50.

[10] J.-J. G. Smart, Sensations and the brain processes, Philosophical Review, 68, 1959, 141–156.

[11] H. Feigl, The "mental" and the "physical", Minn. Studies of Philosophical Science, 2, 370–497, 1958.

[12] N. Block, Troubles with functionalism, Minn. Studies of Philosophical Science, 9, 1978, 261–325.

[13] N. Block, What is functionalism? In: N. Block (ed.), Readings in the Philosophy of Psychology, Cambridge, MA: Harvard University Press, 1980, pp.171–184.

[14] D. Chalmers, The Conscious Mind: In Search of a Fundamental Theory, Oxford, UK: Oxford University Press, 1996.

[15] S. Dehane, C. Sergent and J.P. Changeux, A neuronal network model linking subjective reports and objective physiological data during conscious perception, Proceedings National Academic Science, U.S.A., 2003, 8520–8525.

[16] N. Block, Comparing the major theories of consciousness, In: M. Cazzaninga (ed.), The Cognitive Neurosciences IV (Ch. 77), Cambridge, MA, USA: MIT Press, 2009.

[17] K. Stone, U.T. Place, "Is Consciousness a Brain Process? Philosophy of mind (PartI : Foundations). http://kemstone.com/Nonfiction/Philosophy/Mind/placeidentity.htm

[18] S. Nath, J.J.C. Smart in Defense of Place's identity of mind, IORS J. of Humanities and Social Science, 19(2), 2014, 26–29.

[19] D.C. Dennet, The International Stance, Cambridge, MA: MIT Press, 1987.

[20] L.J. Byom and B. Mutlu, Theory of mind: Mechanisms, methods, and new directions, Neuroscience, 7, 2013, 413–428.

[21] B. Scassellati, Theory of mind... for a robot, AAAI Technical Report FS-00–04, 2000.

[22] A.M. Leslie, Spatiotemporal continuity and the perception of causality in infants, Perception, 11, 1984, 287–305.

[23] A.M. Leslie, ToMM, TOBY, and agency: Core architecture and domain specificity, In: L.A. Hirschfield and S.A. Gelman (eds.), Mapping the Mind: Domain Specificity in Cognition and Culture, Cambridge, U.K.: Cambridge University Press, 1994, 119–148.

[24] S. Baron-Cohen, Mind Blindness, Cambridge, MA: MIT Press, 1995.

[25] J. Weng, J. McCelland, A. Pentland, O. Sporns, I. Stockman, M. Sur and E. Thelen, Autonomous mental development by robots and animals, Science, 291(26), 2001, 599–600.

[26] M. Lungarella, Developmental robotics, http://www.scholarpedia.org/article/Developmental_robotics

[27] A. Borji and L. Itti, State-of-the-art in visual attention modeling, IEEE Transactions on Pattern Analysis and Machine Intelligence, 35(1), 2013, 185–207.

[28] J.F. Ferreira and J. Dias, Attentional mechanisms for socially inter-active robots: A survey, IEEE Transactions on Autonomous Mental Development, 6(2), 2014, 110–125.

[29] R. Bajcsy, Active perception vs. passive perception, Proceedings 3rd IEEE Workshop on Computation, Bellair, Michigan, 1985, 55–59.

[30] G.O. Deak, I. Fase and J. Movellan, The emergence of shared attention: Using robots to test developmental theories, Proceedings of the 1st International Workshop on Epigenetic Robotics, Lund University Cognitive Studies, 1985, 95–104.

[31] C. Sidner and C. Lee, Attentional gestures in dialogues between people and robots, In: T. Nishida (ed.) Engineering Approaches to Conversational Informatics, Ch.6, New York, U.S.A.: Wiley, 2007.

[32] L.-P. Morency, C. Sidner and D. Darrell, Dialogue context for visual feedback recognition, In: T. Nishida (ed.), Ch. 1, New York, U.S.A.: Wiley, 2007.

[33] K. Dautenhahn, Robots we like to live with? A developmental perspective on a personalized life-long robot companion, Proceedings 13th IEEE International Workshop on Robot and Human Interactive Communication (RO-MAN), 2004, 17–22.

[34] K. Dautenhahn, C.I. Nehaniv, B. Walters, B. Robins, H. Kose, N. Bagci, A. Mirza and M. Blow, KASPAR: A minimally expressive humanoid robot for human-robot interaction research, Applied Bionics and Biomechanics, 6(3), 2009, 369–397.

[35] M. Salichs, R. Barber, A. Khamis, M. Malfaz, J. Garostiza, P. Pachero, Maggie: A robotic platform for human-robot social interaction, Proceedings of 2006 IEEE Conference on Robotics, Automation and Mechatronics, Bangock, Thailand, 7–9 June, 2006, 1–7.

[36] T. Kanda, H. Ishiguro, T. Ono, M. Imai and R. Nakatsu, Development and evaluation of an interactive humanoid robot: Robovie, Proceedings of IEEE International Conference on Robotics and Automation, Washington, DC, U.S.A., May 2002, 1848–1855.

[37] E. Stradella, P. Salvini, A. Pinni, A. DiCarlo, C.M. Oddo, P. Dario and E. Palmerini, Robot Companions as case-scenario for assigning the "subjectivity" of autonomous agents: Some philosophical and legal remarks.

[38] H.T. Engelhardt, Some persons are humans, some humans are persons, and the world is what we persons make of it, In. S.F. Spicker and H.T. Engelhardt (eds.), Philosophical medical ethics : Nature and significance, Dordrecht: D. Reidel, 1977.

[39] E. Hilgendorf and M. Kim, Legal regulation of autonomous systems in South Korea on the example of robot legislation, Proceedings of International Workshop on Autonomics and Legal Implications, Berlin, Germany, 2 November, 2012, (http://gccsr.org/node/685).

[40] E. Palese, Robots and cyborgs; to be or to have a body? (Springer Online: May 30, 2012). http://www.nebi.nlm.nih.gov/pmc/articles/PMC 3368120/

[41] M.S. Kumar, Cyborgs: The future mankind, International Journal of Scientific and Engineering Research, 5(5), 2014, 414–420.

[42] Seven real-life human cyborgs (Mother Nature Network: MNN, 26 Jan 2015). http://www.mnn.com/leaderboard/stories/7-real-life-human-cyborgs

[43] J. Andraide, The contribution of working memory to conscious experience, In: J. Andraide, et al. (eds.), London: Psychology Press/Taylor and Francis, 2001.

[44] B.J. Baars and S. Franklin, How conscious experience and working memory interact, TRENDS in Cognitive Sciences, 17(4), 2003, 165–172.

[45] S. Nagataki, M. Shibata, T. Konno, T. Hashimoto and H. Hattori, Joint attention realized in a robot with intentional agency, Proceedings ECC'10: European Control Conference, Lisbon, Portugal, 14, September, 2010.

[46] C.-M. Huang and B, Mutlu, Robot behavior toolkit generating effective social behaviors for robots, Proceedings of 8th ACM/IEEE International Conference on Human-Robot Interaction (HRI'12), Boston, MA, 2012, 25–32.

Index

About the Author

Spyros G. Tzafestas was born in Corfu, Greece on December 3, 1939. Degrees: B.Sc. in Physics (1962) and P.G. Dipl. in Communications (1963) from Athens Univ. D.I.C. and M. Sc. (country-region Eng.) in Control from London Univ. (1967). Ph.D. in Systems and Control from Southampton Univ. (March 1969). Positions: Research leader at 'Demokritos' Res. Center in Computer Control (1969–1973). Professor of Control Systems at the E. Eng. Dept. of Patras Univ., Patras, Greece (Oct. 1973–Mar. 1985). Professor in Control and Robotics, Dir. of R&A Lab. at the School of E.& C. Eng. (SECE) of the Natl. Tech. Univ. of Athens (NTUA, Apr. 1985–Aug. 2006), Dir. of NTUA Inst. Comm. & Comp. Syst. (1999–2009). Supervisor of 35 Ph.D., and over 100 Master Theses. Recipient of D.Sc. (Southampton Univ.) and two honorary doctorates in engineering (TU Munich, Germany, and EC Lille, France). Over the years he worked in D. P.S., M.V. and M.D. Systems, Robotic systems, and AI, KB and CI control techniques (over 700 papers). He is a LIFE FELLOW of IEEE and a FELLOW of IET (IEE). Founding editor of J. Int. and Robotic Syst. (1988–2006). Chief Editor of the Springer ISCA book Series. Editor of 40 research books and 25 journal special issues. Organizer and/or chair of many Intl. conferences (IEEE CDC, EUCA, etc.), and coordinator of many national and European projects in IT, CIM, Robotics, Intelligent Systems, and Control. Author of five international books and seven Greek books, His biography is included in more than 30 intl. biographical volumes. Currently, Dr. Tzafestas continues his scientific work at NTUA as a Professor Emeritus-Senior Research Associate of SECE.

309